2nd International Conference on Functional Materials and Metallurgy (ICFMM 2017)

IOP Conference Series: Materials Science and Engineering
Volume 303

Kuala Lumpur, Malaysia
28 – 30 November 2017

ISBN: 978-1-5108-5688-2
ISSN: 1757-8981

Printed from e-media with permission by:

Curran Associates, Inc.
57 Morehouse Lane
Red Hook, NY 12571

Some format issues inherent in the e-media version may also appear in this print version.

Printed by Curran Associates, Inc. (2018)

Institute of Physics
Dirac House, Temple Back
Bristol BS1 6BE UK

Phone: 44 1 17 929 7481
Fax: 44 1 17 920 0979

techtracking@iop.org

Additional copies of this publication are available from:

Curran Associates, Inc.
57 Morehouse Lane
Red Hook, NY 12571 USA
Phone: 845-758-0400
Fax: 845-758-2633
Email: curran@proceedings.com
Web: www.proceedings.com

Table of contents

Volume 303

2017 the 2nd International Conference on Functional Materials and Metallurgy (ICFMM 2017)

28–30 November 2017, Kuala Lumpur, Malaysia

Accepted papers received: 11 January 2018
Published online: 30 January 2018

Preface

Papers

New Material Design and Performance Analysis

IOP Conf. Series: Materials Science and Engineering **303** (2017) 011001 doi:10.1088/1757-899X/303/1/011001

PREFACE

It is our great pleasure to welcome you to 2017 2nd International Conference on Functional Materials and Metallurgy which has been held in Kuala Lumpur, Malaysia during November 28-30, 2017. ICFMM 2017 is dedicated to issues related to functional materials and metallurgy.

The major goal and feature of the conference is to bring academic scientists, engineers, industry researchers together to exchange and share their experiences and research results, and discuss the practical challenges encountered and the solutions adopted. Professors from Singapore, Australia and China are invited to deliver keynote speeches regarding latest information in their respective expertise areas. It is a golden opportunity for the students, researchers and engineers to interact with the experts and specialists to get their advice or consultation on technical matters, teaching methods and strategies.

These proceedings present a selection from papers submitted to the conference from universities, research institutes and industries. All of the papers were subjected to peer-review by conference committee members and international reviewers. The papers selected depended on their quality and their relevancy to the conference. The volume tends to present to the readers the recent advances in the field of new material design and performance analysis as well as mechanical design and digital manufacturing systems etc.

We would like to thank all the authors who have contributed to this volume and also to the organizing committee, reviewers, speakers, chairpersons, and all the conference participants for their support to ICFMM 2017.

ICFMM 2017
Organizing Committee
Dec 20th, 2017

Committees

Advisory Chair
Prof. Druce Dunne, University of Wollongong, Australia

Conference Chairs
Prof. Dr. Mohd Hamdi Abdul Shukor, University of Malaya, Malaysia
Prof. Puneet Tandon, Indian Institute of Information Technology, Design and Manufacturing Jabalpur, India

Local Chairs
Prof. Wirachman Wisnoe, Universiti Teknologi MARA, Malaysia
Assoc. Prof. Tang S. H., Universiti Putra Malaysia, Malaysia

Program Chair
Prof. Huijun Li, University of Wollongong, Australia
Prof. Dongyun WAN, Shanghai University, China

Technical Committee
Dr. Azzura Ismail, Universiti Tun Hussein Onn Malaysia, Malaysia
Assoc. Prof. Liang Hao, Tianjin University of Science and Technology
Asst. Prof. Belmouden Moustapha, Faculté des Sciences d'Agadir, Morocco
Prof. Alexander Pesin, Nosov Magnitogorsk State Technical University, Russia
Asst. Prof. Sanghun Lee, Gachon University, Korea
Prof. BLEDNOVA Zhesfina Michailovna, Kuban State Technological University, Russia
Sarum Boonmee, Suranaree University of Technology, Thailand
Prof. Afang Zhang, Shanghai University, China
Assoc.Prof. Ke Chen, Shanghai Jiao Tong University, China
Assoc.Prof. Xiang Li, University of Shanghai for Science and Technology, China
Prof. Osman Adiguzel, Firat University, Turkey
Dr. Hao Wang, Northeastern University, China
Lecturer Kevin Alvin Eswar, UNIVERSITI TEKNOLOGI MARA (UiTM), Malaysia
Assoc Prof. WangJing Ma, The Chinese Academy of Sciences, China
Asst. Prof. Naveen B., SDM College of Engineering & Technology, India
Dr. Lomas Tomar, University of the Witwatersrand,South Africa
Assoc. Prof. A.Mathiazhagan, Cochin University, India
Prof. P Sarah,Vardhaman College of Engineering, India
Lecturer Attasit Sirivachiraporn, Srinakharinwirot University, Thailand
Asst.Prof. Ranjith G Nair, National Institute of Technology Silchar, India

IOP Conf. Series: Materials Science and Engineering **303** (2017) 011003 doi:10.1088/1757-899X/303/1/011003

Peer review statement

All papers published in this volume of *IOP Conference Series: Materials Science and Engineering* have been peer reviewed through processes administered by the proceedings Editors. Reviews were conducted by expert referees to the professional and scientific standards expected of a proceedings journal published by IOP Publishing.

IOP Conf. Series: Materials Science and Engineering **303** (2017) 012001 doi:10.1088/1757-899X/303/1/012001

Performance engineering of dye sensitized solar cells (DSSC) using Ag modified titania as photoanode

Ranjith G. Nair[1,a], Mathan Kumar P.[1,b] and S. K. Samdarshi[2,c]

[1]Solar Energy Materials Research and Testing Laboratory (SMaRT Lab), Department of Physics, National Institute of Technology Silchar, Cachar, Assam-788010, India.
[2] Centre for Energy Engineering and Centre of Excellence in Green and Efficient Energy Technology (CoE-GEET), Central University of Jharkhand, Ranchi, 835205, India

a rgnair2007@gmail.com, b mathannits92@gmail.com,
c drsksamdarshi@rediffmail.com

Abstract. Present work reports the fabrication of silver (Ag) modified titania photoanode as an efficient photoanode for Dye Sensitized Solar Cell (DSSC). Pristine and Ag modified Titania nanomaterials were prepared using sol gel method. The structural analyses confirm the high crystallinity of the samples with crystallite size distribution in nanorange. TEM micrograph confirms that the synthesized nanomaterials are in uniform size. A red shift is observed in the UV DRS spectra compared to pristine Titania and which confirm the incorporation of Ag inside titania. A prototype DSSC was fabricated using the pristine and modified Titania as photoanode, Ruthenium dye as sensitizer, I-/I-3 as redox electrolyte and platinum counter electrode. The cell with Ag modified titania photoanode showed 15 times enhanced photoconversion efficiency (PCE) than the pristine one. This improved performance of the Ag modified DSSC can be ascribed to reduced recombination and improved charge carrier transport of electrons/ holes at the interfaces.

1. Introduction

Dye-sensitized solar cells (DSSCs) is a promising alternative to the traditional silicon photovoltaic technology due to its low cost, flexibility, and easy fabrication[1]. The maximum reported photoconversion efficiency of DSSC is 13%[2]. Stability, interfacial charge carrier transport and recombination of charge carriers are the major factors that limit DSSC for commercialization. The performance of the DSSC can be improved by increasing the light harvesting, reducing interfacial charge carrier recombination and increasing the charge carrier transport at the interfaces [3].

Multiple attempts were made over decades to solve the above issues through modified photoanodes[4], increased dye loading and tuning its absorption[5], replacing liquid to solid state electrolyte[6] and modifying the interfaces to favor the electron transport[7]. Recently it has been reported that, incorporation of noble metals such as gold, silver etc. in photoanode can improve the photoconversion efficiency through tuning the dye absorption coefficient due to Surface Plasmon Resaonance (SPR), reducing interfacial recombination and improving charge carrier transport[8-11].

Based on the above present work investigates the synthesis and characterization of Ag modified titania as an effective photoanode for DSSC and its photovoltaic performance. Initially pristine and Ag modified Titania nanoparticles were synthesized and its physico-chemical characteristics were studied.

IOP Conf. Series: Materials Science and Engineering **303** (2017) 012001 doi:10.1088/1757-899X/303/1/012001

A prototype Dye Sensitized Solar Cells were fabricated using pristine and Ag modified Titania as photoanode. The photovoltaic performances of the fabricated DSSC were studied using I-V characteristics and compared. The open circuit voltage and the short circuit current are measured under light and dark conditions.

2. Experimental

2.1. Materials
The precursors used for the synthesis of Ag modified Titania were Titanium iso-propoxide with 95% purity from Sigma Aldrich [Ti (OCH (CH3)2)4], isopropyl alcohol [CH3)2(CHOH] and Silver nitrate (AgNO3) from Merck. FTO glass substrates of 2.2mm thick with a surface resistivity of 7 Ω cm-2 with 80% transmission, Ruthenizer 535-bisTBA or N719 , Platisol T (a liquid paint containing a chemical platinum precursor) and Iodolyte AN-50 were obtained from Solaronix SA, Switzerland.

Pristine Titania was synthesized by mixing titanium iso-propoxide with propanol in 1:10 ratio at room temperature under constant stirring to form sol and kept 24 hrs to obtain gel. Ag modified Titania was prepared by adding 2.5 wt.% of AgNO3 solution in 1ml of water to the Titania sol at room temperature under constant stirring to form resultant sol. The substrates were cleaned with ethanol, acetone and sonication followed by washing with double distilled water. The pristine and Ag modified Titania thin films were prepared by gently dipping the substrates into the as prepared and Ag modified sol respectively. The films were then dried at 100oC for 10 min. This process is repeated to get the desired thickness. The prepared films were then annealed at 400°C for 1 hour.

The N719 dye solution was prepared by mixing 10mg of dry dye powder in 25mL of methanol The prepared Ag modified photoanode was then cooled down and soaked into the dye solution for 2 days in a sealed beaker at room temperature for dye sensitization. The sensitized photoanode is washed with ethanol to remove the additional dye absorbed portions and it was dried at 60°C. The platinum counter electrode was prepared by gently painting the platinum precursor ink with a regular brush on FTO substrate. Later the platinum-coated conducting glass (counter electrode) is sandwiched with N719 dye-sensitized electrode (working electrode) with clips. A proper amount of redox electrolyte, Iodolyte AN-50(Solaronix SA, Switzerland) was injected in between the two glass plates and gently press the clips to allow the electrolyte to spread along the surface Finally the device was sealed properly to avoid electrolyte leakage.

2.2. Characterization
The structural analyses of the prepared samples were studied using X ray diffraction (XRD) pattern, with a diffractometer (Rigaku Miniflux,Japan) using Cu Kα radiation (λ=1.5Å) at 30 kV and 450W at a scan rate of about 0.05° 2θ per second and scan range between 10-70o and Transmission Electron Microscope (TEM) recorded using Joel Jem- 1000CX II, Japan. The surface and elemental analyses of the samples were recorded using Scanning electron microscopy (SEM) employing scanning electron microscope (Joel 6390 LV,Japan) operated at an operating voltage of 15 kW and Energy dispersive X-ray analysis(EDAX) using the EDAX (Oxford instrument, UK) attached to the SEM system. The optical properties of the samples were studied using Shimadzu UV-2200, Japan UV-visible spectrophotometer with diffuse reflectance attachment for UV- Visible light response and BaSO4 was used as reference. The solar cell performance of the samples were studied using I-V characteristics with Potentiostat /Galvanostat (AUTOLAB 302N, Metrohm, Holland). The performance of the cell was tested under both dark and white light illumination (100 W/m2 at 1.5 AM).

3. Results and Discussion
The XRD spectra of Ag modified Titania calcined at 400°C is shown in Figure 1. The spectra show that Ag modified Titania is in anatase phase with nanoscale crystallite distribution and which is in good agreement with the JCPDS (card no 73-1764) data.

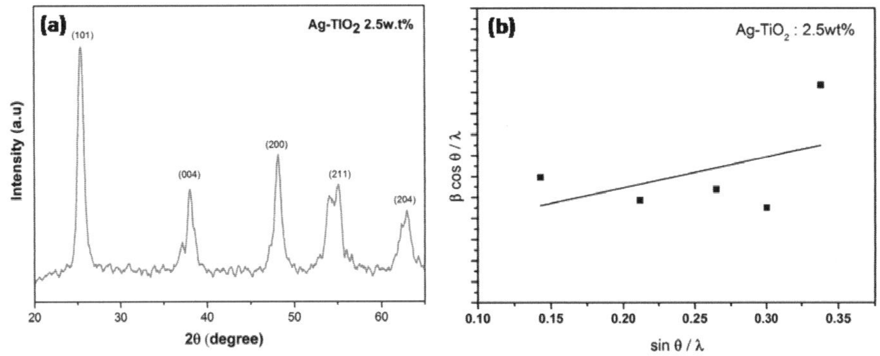

Figure 1. XRD Spectra (a) and W-H Plot (b) of Ag modified Titania

A peak shift is observed in Ag modified Titania compared to pristine Titania and which can be ascribed to the incorporation of Ag inside the Titania matrix. The crystallite size of the samples were calculated using Scherrer's equation[12]. The effective crystallite size of the samples were calculated without strain using Williamson Hall equation[13] The WH plot of Ag modified Titania shows the positive slope and which confirms the presence of tensile strain. The intercept values give the crystallite size without strain, which was found to be 18.10 nm for Ag modified Titania. The specific surface area was also calculated using the formula S= 6/pD, where p is the material density and D is the crystallite size obtained from Scherer's equation. The values of the crystallite size, strain and specific surface area are given in Table 1.

Table 1. Physicochemical characteristics of the samples.

Sample	Strain	Crystallite size (nm)			Specific surface area (m^2/g)	Bang Gap (eV)	EDS (Atom %)	
		Scherrer	WH	TEM			Ag	Ti
Ag-TiO_2	0.0147	10.5	18.18	12.5	145.793	2.7	5.48	94.52

Transmission Electron Microscope (TEM) analyses were carried out to understand the structure of the samples and shown in Figure-2a. The micrograph confirms that the crystallite size of the samples was in nanorange (~12.5nm) and which is in accordance with the XRD data.

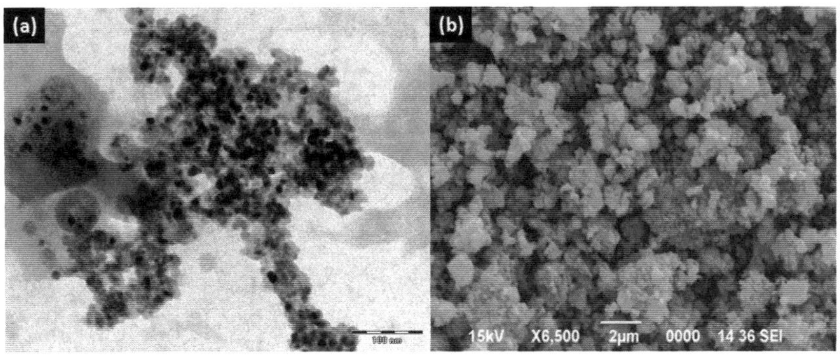

Figure 2. TEM (a) and SEM (b) images of Ag modified Titania

Surface and elemental analyses of the sample was carried out using Scanning Electron Microscopy and Energy Dispersive analyses respectively and shown in Figure-2b. The SEM micrographs of the sample revels that the particles are highly agglomerated. The EDAX analyses of the sample confirm the Ag loading inside the Titania matrix and shown in Table 1.

IOP Conf. Series: Materials Science and Engineering **303** (2017) 012001 doi:10.1088/1757-899X/303/1/012001

UV-Visible spectroscopy of the samples is shown in figure 3. The spectrum shows the absorption edge of pristine Titania to be 385nm [14] and it is in agreement with the XRD results. A noticeable red shift is observed in the Ag modified Titania which can be attributed to the incorporation of Ag inside the Tiatnia matrix[15]. The average band gap of the sample was calculated from the Tauc plot and the values are summarized in the Table 1.

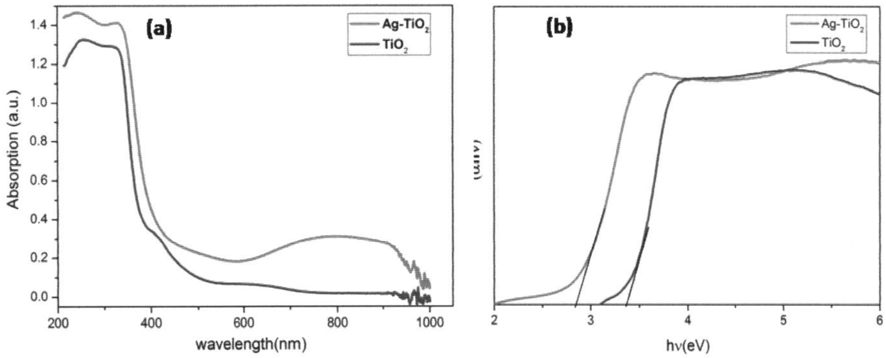

Figure 3. UV DRS Spectra (a) of Ag modified Titania along with Tauc,s Plot (b)

The current – voltage characterization of the fabricated solar cell was measured using 500W halogen lamp (Philips Halotone plus Line). The photovoltaic performance of the pristine and Ag doped cells were tested under both dark and white light illumination. The I-V characteristics of the fabricated cell are shown in Figure 4 and the values of Short circuit current, open circuit voltage, fill factor and efficiency are summarized in Table 2. The addition of Ag content greatly reduces the recombination and facilitates the photo generated electrons from the dye to the Titania due to the creation of electron-hole separation centers at the Ag / TiO2 interface and thus increases the Short circuit current jsc. Whereas the significant improvement in the Open circuit voltage Voc is due to the shift of quasi-Fermi level of Ag-TiO2 to more negative level compared to the Pristine TiO2[11]. The photo conversion efficiency of the Ag/TiO2 cell is found to be 15 times more than the Pristine TiO2 cell.

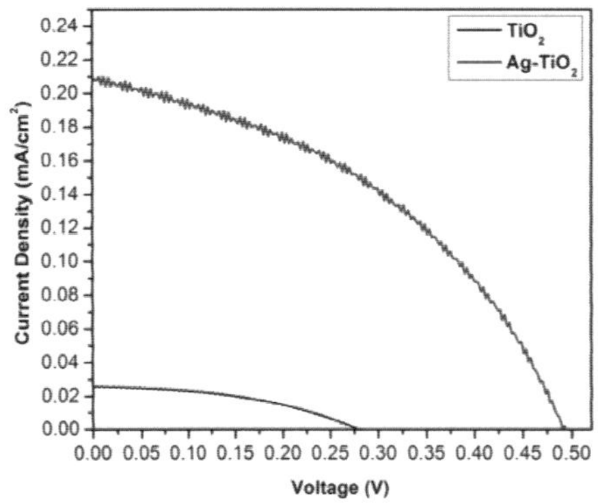

Figure 4. J-V characteristic of the fabricated cells

Table 2. Photovoltaic performance of Pristine and Ag Modified Titania

Electrode	V_{oc} (V)	J_{sc} (mA)	V_{max} (V)	J_{max} (mA)	FF	η (%)
TiO_2	0.277	1.374	0.147	0.957	0.371	0.028
Ag/TiO_2	0.490	10.600	0.302	7.040	0.409	0.421

4. Conclusion

Pristine and Ag modified titania photoanodes were prepared using a modified sol-gel method. The structural, surface, elemental and optical characterization of the samples were analyzed using XRD, TEM, SEM, EDS and UV-Vis spectroscopy. The structural analyses confirm the pristine anatase phase of the sample with nanoscale distribution. Surface analysis reveals that the particles are uniformly distributed with little agglomeration. Elemental analysis along with UV-visible spectra confirm the presence of Ag in the sample. A Dye Sensitized Solar Cells (DSSC) was fabricated using pristine and modified titania as photoanode, and its performance were tested under light and dark conditions. J-V characteristic of the Ag modified photoanode based DSSC shows an increased short circuit current, open circuit voltage, fill factor and photoconversion efficiency compared to pristine photoanode. The photo conversion efficiency of the Ag/TiO2 cell is found to be 15 times more than the pristine TiO2 cell. This improved performance is due to incorporation of Ag which reduces the interfacial recombination and enhances the transport of electron at the interfaces. The performance of the fabricated cell may be further improved by achieving surface plasmon resonance in the photoanode.

References

[1] M. Grätzel, "Dye-sensitized solar cells," J. Photochem. Photobiol. C Photohem. vol. 4, pp. 145–153, 2003.

[2] S. Mathew, A. Yella, P. Gao , R. H. Baker, B. F. E. Curchod, N. A. Astani, I. Tavernelli, U. Rothlisberger, Md. K. Nazeeruddin, M. Gra¨tzel1 "Dye-sensitized solar cells with 13% efficiency achieved through the molecular engineering of porphyrin sensitizers,", Nat. Chem. pp. 1–6, 2014.

[3] B. Z. Wang, Y. Cui, K. Hara, Y. Dan-oh, C. Kasada, and A. Shinpo, "A High-Light-Harvesting-Efficiency Coumarin Dye for Stable Dye-Sensitized Solar Cells ," Adv.Mater, pp. 1138–1141, 2007.

[4] Q. Sun, Y. Li, J. Dou, and M. Wei, "Improving the efficiency of dye-sensitized solar cells by photoanode surface modifications," Sci China Mater, vol. 59, no. 10, pp. 867–883, 2016.

[5] T. P. Chou, Q. Zhang, G. Cao, M. Science, V. Uni, and R. Hall, "Effects of Dye Loading Conditions on the Energy Conversion Efficiency of ZnO and TiO 2 Dye-Sensitized Solar Cells," J. Phys. Chem. C, pp. 18804–18811, 2007.

[6] I. Chung, B. Lee, J. He, R. P. H. Chang, and M. G. Kanatzidis, "All-solid-state dye-sensitized solar cells with high efficiency," *Nature*, vol. 485, no. 7399, pp. 486–489, 2012.

[7] A. H. G. Niaki, A. M. Bakhshayesh, and M. R. Mohammadi, "Double-layer dye-sensitized solar cells based on Zn-doped TiO 2 transparent and light scattering layers : Improving electron injection and light scattering effect," *Sol. Energy*, vol. 103, pp. 210–222, 2014.

[8] S. P. Lim, Y. S. Lim, A. Pandikumar, H. N. Lim, Y. H. Ng, R. Ramaraj, D. C. Sheng Bien, O. K. Abou-Zied, N. M. Huang. "Gold–silver@TiO_2 nanocomposite-modified plasmonic photoanodes for higher efficiency dye-sensitized solar cells," *Phys. Chem. Chem. Phys.*, vol. 19, no. 2, pp. 1395–1407, 2017.

[9] *H. Hu, J. Shen, X. Cao, H. Wang, H. Lv , Y. Zhang, W. Zhu, J. Zhao, C. Cui,.*, "Photo-assisted deposition of Ag nanoparticles on branched TiO2 nanorod arrays for dye-sensitized solar cells with enhanced efficiency," *J. Alloys Compd.*, vol. 694, pp. 653–661, 2017.

[10] D. Eli, M. Y. Onimisi, S. G. Abdu, P. M. Gyuk, and E. Jonathan, "Enhanced Performance of a Dye Sensitized Solar Cell Using Silver Nanoparticles Modified Photoanode," *JSRR*, vol. 10, no. 4, pp. 1–8, 2016.

[11] K. Guo, M. Li, X. Fang, X. Liu, B. Sebo, Y. Zhu, Z. Hu, X. Zhao "Preparation and enhanced properties of dye-sensitized solar cells by surface plasmon resonance of Ag nanoparticles in nanocomposite photoanode," *J. Power Sources*, vol. 230, pp. 155–160, 2013.

[12] R. G. Nair, S. Paul, and S. K. Samdarshi, "High UV/visible light activity of mixed phase titania: A generic mechanism," *Sol. Energy Mater. Sol. Cells*, vol. 95, no. 7, pp. 1901–1907, 2011.

[13] V. Mote, Y. Purushotham, and B. Dole, "Williamson-Hall analysis in estimation of lattice strain

in nanometer-sized ZnO particles," *J. Theor. Appl. Phys.*, vol. 6, no. 1, p. 6, 2012.
[14] G. M. Kim, S. M. Lee, G. H. Michler, H. Roggendorf, U. G sele, and M. Knez, "Nanostructured pure anatase titania tubes replicated from electrospun polymer fiber templates by atomic layer deposition," *Chem. Mater.*, vol. 20, no. 9, pp. 3085–3091, 2008.
[15] S. Sun, J. Bao, C. Gao, and J. Ding, "Photocatalytic degradation of gaseous o-xylene over M-TiO 2 (M=Ag, Fe, Cu, Co) in different humidity levels under visible-light irradiation: Activity and kinetic study," *Rare Met.*, vol. 30, no. SUPPL.1, pp. 147–152, 2011.

Acknowledgement
The financial supports provided by the Inter University Acceleration Centre, New Delhi vide Project no. UFR-58315 and DST-SERB, New Delhi vide Project no. ECR/2016/001186 is gratefully acknowledged.

IOP Conf. Series: Materials Science and Engineering **303** (2017) 012002 doi:10.1088/1757-899X/303/1/012002

Tailoring perovskite compounds for broadband light absorption

Hengchang Lu, Xiaowei Guo* , Cheng Yang and Shaorong Li

School of Optoelectronic Information, University of Electronic Science and Technology of China, Chengdu, 610054, China

gxw@uestc.edu.cn

Abstract. Perovskite solar cells have experienced an outstanding advance in power conversion efficiency (PCE) by optimizing the perovskite layer morphology, composition, interfaces, and charge collection efficiency. To enhance PCE, the mixed perovskites were proposed in recent years. In this study, optoelectronic performance of pure perovskites and mixed ones were investigated. It was demonstrated that the mixed perovskites exhibit superior to the pure ones. The mixed material can absorb broadband light absorption and result in increased short circuit current density and power conversion efficiency.

1. Introduction

Since the pioneering work of Snaith's group[1], who demonstrated a simple planar heterojunction solar cell, planar perovskite devices have been receiving great attention due to the simplicity in processing[2]. A lot of work was conducted including perovskite compound engineering, hole and electronic transfer material engineering, interface engineering, and device structure engineering. Thanks to past efforts, an efficiency of up to 22.1% for perovskite solar cell was reached in 2016[3], superior to their organic rivals and comparable to silicon solar cell.

It has already showed that chemical modification of cations[4] and anions[5] in perovskite materials can tune the bandgaps. For example, substitution of the anion I with Br in $MAPbI_3$ (MA= methylammonium) results in the bandgap variation from 1.5 eV to 2.3 eV. On the other hand, $HC(NH_2)_2PbI_3$,which contains formamidinium (FA) cations instead of MA cations in the perovskite structure, has a bandgap of 1.48 eV, with an absorption edge of 840 nm. MA-based lead iodide ($MAPbI_3$) perovskite materials possess a band gap of 1.57 eV. A strategy for extending the absorption range of solar light is thus to replace $MAPbI_3$ with $FAPbI_3$. Replacing the MA cation with the FA cation, which has an ionic radius relatively larger than that of MA^+ in the organic-inorganic lead iodide perovskite material, can change the metal-halide-metal bond angle and thus leads to a narrower band gap compared to that of $MAPbI_3$. However, $FAPbI_3$ has two crystal structures[5], i.e., a trigonal structure (perovskite phase, black color, α-$FAPbI_3$) and a hexagonal structure (nonperovskite phase, yellow color, δ-$FAPbI_3$), depending on the synthesis temperature. The α-$FAPbI_3$ is stable at a high temperature (over 160 °C) and can convert into δ-$FAPbI_3$ under an ambient humid atmosphere. To prevent the transformation of α-$FAPbI_3$ to δ-$FAPbI_3$, preparation of $(FAPbI_3)_{1-x}(MAPbI_3)_x$, which is a more stable new perovskite, was regarded as an effective route.

Here, a comparison of optoelectronic performance between pure and mixed perovskites were reported , including morphology and absorption spectrum as well as the device performances of different perovskite solar cells.

2. Experimental Section

2.1 Synthesis of the inorganic–organic mixed perovskite
Methylammonium iodide (CH_3NH_3I) and formamidinium iodide ($NH_2CH_5NH_2I$) were synthesized according to the reported procedure[7,8] by reacting the hydroiodic acid (10 mL, 57 wt % in water, Sigma-Aldrich) and methylamine and mormamid (24 mL, 33 wt % in absolute ethanol, Sigma-Aldrich) at 0 °C with stirring for 60 min. Raw CH_3NH_3I and $NH_2CH_5NH_2I$ were obtained by removing the solvent at 50 °C on a rotary evaporator. The material was then washed in diethyl ether and filtered several times. The precipitate was then dried in vacuum oven overnight at 50 °C and then kept in nitrogen-filled glovebox.

The perovskite solution was prepared by mixing 1.2 mmol of PbI_2 (or $PbBr_2$) (99.99%, Alfa Aesar) and 1.2 mmol of CH_3NH_3I(or $NH_2CH_5NH_2I$) in 1 mL of anhydrous N,N-dimethylformamide (DMF, Sigma-Aldrich). To prepare perovskite films, 50 μL of perovskite precursor solutions was spin coated onto the substrates at 5000 rpm for 55 s. During the spin-coating, 50 μL of anhydrous chlorobenzene (Sigma-Aldrich) was quickly dropped in the center of the substrates after six or seven seconds. The spin-coated films were transferred onto a hot plate for drying at 100 °C for 2 min and then annealed under different atmospheres at 100 °C for 30 min. The desired solutions of $MAPbI_3$, $FAPbI_3$, $MAPbBr_3$, $FAPbBr_3$, $(FAPbI_3)_{1-x}(MAPbI_3)_x$, and $(FAPbBr_3)_{1-x}(MAPbBr_3)_x$ were prepared.

2.2 Device fabrication
Solar cells with ITO/PEDOT:PSS/ perovskite/PCBM/Ag planar device structure were prepared. The ITO glass substrates (1.5 cm × 1.5 cm) were ultrasonically cleaned with detergent, deionized water, acetone, and ethanol, sequentially, and then were blow-dried in nitrogen. Before deposition of other films, the ITO substrates were treated with ultraviolet-ozone for 10 min. After that, PEDOT:PSS solution (Clevious Al 4083) was spin-coated on ITO-coated substrates at 3000 rpm for 45 s, and subsequently the PEDOT:PSS films were annealed at 150 °C for 20 min. After perovskite films were deposited, 20 mg/mL of PCBM (FEM. Inc.) solution in chlorobenzene was then spin-coated at 1000 rpm for 45 s. Finally, Ag top contact was deposited by evaporation through an aperture mask under a base pressure of 7×10^{-6} Torr. The active device area is 0.12 cm^2.

2.3 Mesurement and characterization
Scanning electron microscope (SEM) images were obtained using Hitachi S5200. Atomic force microscopy (AFM) measurements were performed using Asylum Research MFP-3D. Current–voltage measurements were carried out using Keithley 2400 at room temperature under AM 1.5G illuminations (1000 W/m^2) from a solar simulator, which was calibrated using a standard silicon solar cell device.

3. Results and discussions

3.1 Morphology of perovskite films
The mixing ratio x was optimized by conducting a series of contrastive experiments. It was found that the performance of mixed perovskites can reach highest at x=0.15, which agrees well with the reorpted[4]. Therefore, $(FAPbBr_3)_{0.15}(MAPbBr_3)_{0.85}$ and $(FAPbI_3)_{0.15}(MAPbI_3)_{0.85}$ were prepared for comparison here. Figure 1 shows the morphology of as-prepared perovskite films on ITO glass. It is found that the perovskites with Br anions have large surface roughness, which can be seen from Figure 1(a),(c),(e). It is interesting that the perovskites with FA cations can produce small roughness. The maximum height of surface roughness on FA perovskites is no more than 70nm, as compared to that one of over 200nm on MA perovskites. In experiments it was obervesed that MA crystalline perovskites with large grain size are easily formed. Therefore, addition of FA perovskites is also beneficial to smooth the device surface, which results in better performance. It is worth pointing out that the rough surface can be made much smoother by post washing or vapour annealing methods[7].

IOP Conf. Series: Materials Science and Engineering **303** (2017) 012002 doi:10.1088/1757-899X/303/1/012002

3.2 Absorption spectrum of perovskite films

Figure 2 presents the absorption spectrum of the perovskite films on ITO glass. For the perovskites with Br anions, it is obvious that the perovskite films show a small absorption edge of about 550nm, corresponding to a band gap of 1.8ev, which agrees well with the reported results in ref.[6]. For the mixed perovskite of $(FAPbBr_3)_{0.15}(MAPbBr_3)_{0.85}$, the absorption behaves worse than that of pure $MAPbBr_3$ but better than that of $FAPbBr_3$. The reason for such low absorption for Br based perovskites is that its large band gap blocks most of the light, which surface shows yellow transparency(Figure 3a). For example, $MAPbBr_3$ has a band gap of about 2.3eV. Therefore, it is very important for high efficiency solar cells to choose the mixing ratio in $(FAPbBr_3)_x(MAPbBr_3)_{1-x}$.

The perovskites with I anions show an absorption edge of about 800nm. As compared to pure perovskites, the mixed ones enhance all the light absorption from 500nm to 1100nm wavelengths. Even in the wavelength range of 800nm to 1100nm the perovskites are not sensitive to, the light absorptions improve greatly. At 900nm, the absorption is enhanced by 3 times and 15 times compared with $FAPbI_3$ and $MAPbI_3$, respectively.

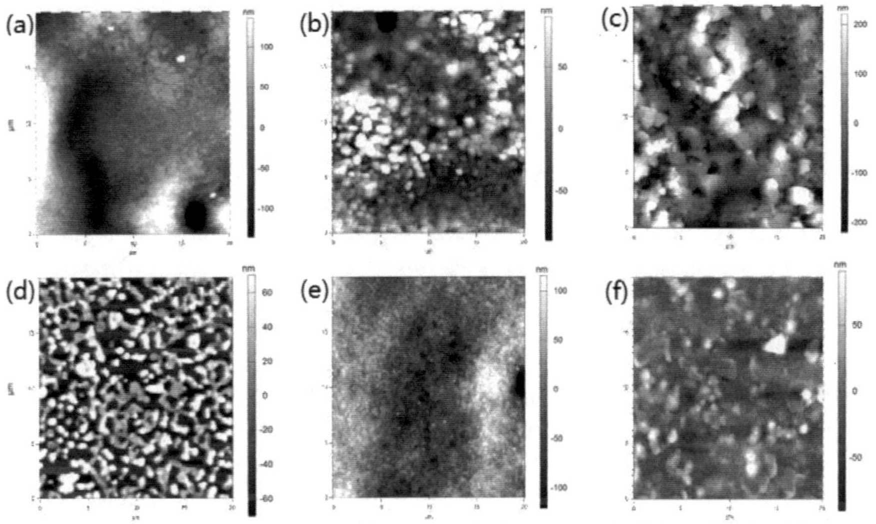

Figure 1. AFM images of (a) $MAPbI_3$, (b) $FAPbI_3$, (c) $MAPbBr_3$, (d) $FAPbBr_3$, (e) $(MAPbBr_3)_{0.85}(FAPbBr_3)_{0.15}$, and (f) $(MAPbI_3)_{0.85}(FAPbI_3)_{0.15}$

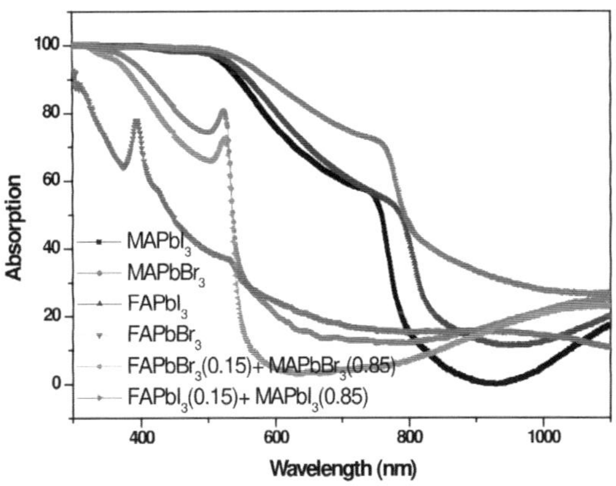

Figure 2. UV-VIS absorption spectrum of all the samples

IOP Conf. Series: Materials Science and Engineering **303** (2017) 012002 doi:10.1088/1757-899X/303/1/012002

Figure 3. The as-prepared perovskites on ITO glass. (a) $MAPbBr_3$; (b) From left to right: $MAPbI_3$ $FAPbI_3$, $(MAPbI_3)_{0.85}(FAPbI_3)_{0.15}$

3.3 *Cell performances*

Figure 4a shows the cross-sectional SEM image of a completed ITO/ PEDOT:PSS/ $(MAPbI_3)_{0.85}(FAPbI_3)_{0.15}$ / PCBM /Au planar solar cell. Each layer could be clearly defined according to the contrast variation of the cross sectional image. It can be seen that the mixed perovskites are porous. This is because we didn't use any post-treatments, such as non-polar solvent washing, thermal annealing, solvent annealing, and so on. Figure 4b shows the current density-voltage (J-V) curves for these cells. The photovoltaic parameters are also summarized in Table 1. The devices with $MAPbBr_3$, $FAPbI_3$, $FAPbBr_3$ present small open circuit voltages (V_{oc}) below 800mV, small short circuit current densities (J_{sc}) below 2.5 mA/cm^2 and a fill factor (FF) of 50%-60%, leading to very low PCEs of no more than 1.1%. As a result, the cell based on the mixed perovskite of $(MAPbBr_3)_{0.15}(FAPbBr_3)_{0.85}$ is naturally deserved with a low PCE. But it is obvious that the cell performance of the mixed perovskites is superior to those with pure materials, which should be attributed to different band gaps of both perovskite materials. For $MAPbI_3$, the performance is improved with a V_{oc} of 853.57mV and a J_{sc} of 9.42 mA/cm^2 although with a relatively low FF of 50.54, and its efficiency increases to 4.06%. When mixing the material with $FAPbI_3$, the V_{oc}, J_{sc}, and FF are all improved and thus PCE continue to increase up to 6.55%, which also accords well to the absorption spectrum illustrated in Figure 2. It was reported that a longer charge diffusion length and superior stability in $FAPbI_3$ films compared to those in $MAPbI_3$[4]. Although our experiments didn't demonstrate this point clearly, $MAPbI_3$ and $FAPbI_3$ mixed perovskites definitely pose a broadband light absorption and consequently deliver a greater efficiency.

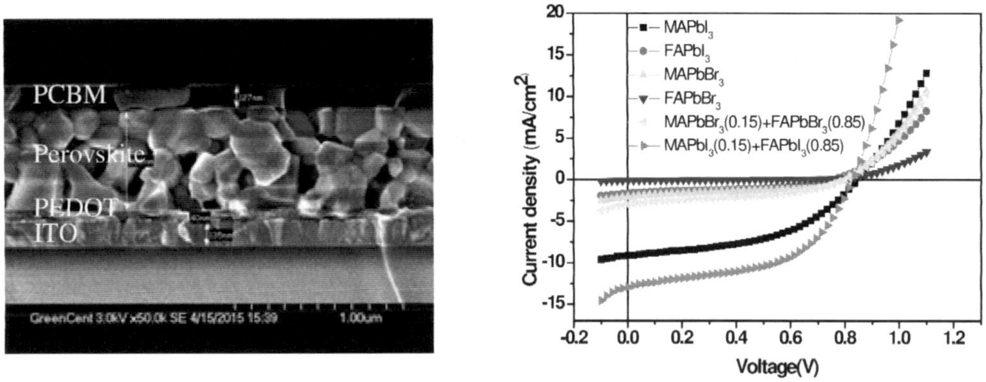

Figure 4. (a) Cross-sectional SEM image of a completed ITO/ PEDOT:PSS/ $(MAPbI_3)_{0.85}(FAPbI_3)_{0.15}$ / PCBM /Au planar solar cell. (b) J–V curves of planar perovskite solar cells with different device structures.

Table 1. Photovoltaic parameters from $(MAPbI_3)_{0.85}(FAPbI_3)_{0.15}$.

Perovskites	V_{oc}	J_{sc} (mA/cm^2)	FF	PCE(%)
$MAPbI_3$	853.57	9.42	50.54	4.06
$FAPbI_3$	746.5	1.76	53.04	0.70
$MAPbBr_3$	784.52	2.3	60.16	1.08
$FAPbBr_3$	705.09	0.15	52.99	0.06
$(MAPbBr_3)_{0.15}(FAPbBr_3)_{0.85}$	862.04	2.48	68.93	1.47
$(MAPbI_3)_{0.15}(FAPbI_3)_{0.85}$	831.71	12.66	62.21	6.55

4. Conclusions

The mixed perovskite of $(MAPbI_3)_{0.85}(FAPbI_3)_{0.15}$ for broadband light absorption have been described. The mixed perovskite of $(MAPbI_3)_{0.85}(FAPbI_3)_{0.15}$ delivers higher performance than pure materials such as $MAPbI_3$, $FAPbI_3$, $MAPbBr_3$, $FAPbBr_3$, and even than other mixed material of $(MAPbBr_3)_{0.15}(FAPbBr_3)_{0.85}$. This mixing strategy may lead to more efficient and cost-effective inorganic-organic hybrid perovskite solar cells.

References

[1] Liu M Z, Johnston M B and Snaith H J 2013, *Nature* **501**,395
[2] Liang J, Wang C X, Wang Y R, Zhaoran Xu Z R, Lu Z P, Ma Y, Zhu H F, Hu Y, Xiao C C, Yi X, Zhu G Y, Lv H L, Ma L B, Chen T , Tie Z X, Jin Z and Liu J 2017, *J. Am. Chem. Soc.*, **139** (7), 2852.
[3] National Renewable Energy Laboratory (NREL) Best Research-Cell Efficiency chart. 2016, http://www.nrel.gov/ncpv/images/efficiency_chart.jpg.
[4] Jeon N J, Noh J H, Yang W S, Kim Y C, Ryu S, Seo J W and Seok S 2015, *Nature* **517,** 476
[5] Eperon G E, Beck C E and Snaith H J 2016, *Mater. Horiz.* **3** (1), 63
[6] Koh T M, Fu K, Fang Y, Chen S, Sum T, Mathews N, Mhaisalkar S G, Boix P P and Baikie T 2014, *J. Phys. Chem. C* **118** (30), 16458
[7] Lee M M, Teuscher J, Miyasaka T, Murakami T N, Snaith H J 2012, *Science* **338**, 643
[8] Liu J, Gao C, He X L, Ye Q Y, Ouyang L Q, Zhuang D M, Liao C, Mei J and Lau W M 2015, *ACS Appl. Mater. Interfaces* **7**, 24008

Acknowledgments

Authors acknowledge the technical discussions of Dr. Jiang Liu from Chengdu Green Energy and Green Manufacturing Technology R&D Centre. The authors also thank finacial support from the Sichuan Science and Technology Project (no. 2016GZ0399).

IOP Conf. Series: Materials Science and Engineering **303** (2017) 012003 doi:10.1088/1757-899X/303/1/012003

Study on ablation behavior of silicone rubber based insulation material under the condition of boron oxide particles erosion

B L Zha, Y A Shi, J J Wang and Q D Su

Xi'an Hi-Tech Institute, Xi'an, 710025, China

shiyiang@gmail.com

Abstract. Self-designed oxygen-kerosene ablation system was employed to study the ablation characteristics of silicone rubber based thermal insulation materials under the condition of boron oxide particles erosion. The ablation test was designed with a mass fraction of 1.69% boron oxide particles and particles-free, the microstructure and elemental analysis of the specimens before and after ablation were carried out by Scanning Electron Microscopy (SEM) and Energy Dispersion Spectrum (EDS). Experiment results show that the average mass ablation rate of the materials was 0.0099 $g \cdot s^{-1}$ and the average ablation rate was -0.025 $mm \cdot s^{-1}$ under the condition of pure gas phase ablation; and the average mass ablation rate of the multiphase ablation test group was 0.1775 $g \cdot s^{-1}$, whose average ablation rate was 0.437 $mm \cdot s^{-1}$; during the ablation process, the boron oxide particles would adhere a molten layer on the flame contact surface of the specimen, which covering the pores on the material surface, blocking the infiltration channel for the oxidizing component and slowing down the oxidation loss rate of the material below the surface, but because the particles erosion was the main reason for material depletion, the combined effect of the above both led to the upward material ablation rates of Silicone Rubber.

1. Introduction

When solid rocket ramjet engine works, the secondary combustion chamber wall should not only endure the ablation of high temperature and oxygen enriched gas, but also suffer from the erosion and damage of metal particles in multiphase flow gas, which puts forward higher requirements for the ablation/erosion properties of insulation materials. Silicone rubber insulation material is widely used as solid propellant rocket ramjet combustion chamber thermal insulation material because of its better oxygen resistance ablation performance, low manufacturing cost and low thermal conductivity.

A large number of ablation tests had been carried out on the ablation process of silicone rubber at home and abroad in order to study the failure mechanism of silicon rubber based insulation materials. Liu Yang and others studied the ablation characteristics of silicone rubber under the condition of particles erosion by means of bending pipe simulation test motor; Wang Shuxian et al. tested the effect of multiphase flow on the ablative properties of silicone rubber insulation. In all kinds of two-phase flow ablation experiments carried out both at home and abroad, most of the particles used in erosion research are dominated by Al_2O_3 particles, Although boron based fuel rich propellants have attracted more and more attention due to their higher calorific value and specific impulse, so far, no boron based oxide ablation/erosion simulation test had been reported. Therefore, it is of great significance to study the ablation behavior of silicone rubber based insulation material with the particles erosion of boron oxide.

In this paper, the oxygen-kerosene ablation system was used to study silicone rubber based insulation material ablation properties under oxygen enriched condition with particles erosion of boron oxide, after the ablation test, the ablation rates were calculated and then combined with the macro and micro morphology of ablation to analyze the influence mechanism of boron oxide. particles on silicon rubber material, the results provided a reference for the study of the failure mechanism of silicone rubber insulation material used to solid rocket ramjet engine using boron propellant.

2. Experimental

2.1. Materials

The raw materials include PDMS; PMPS SiC; fumed silica (silicon dioxide); ammonium polyphosphate (APP); montmorillonite; carbon fiber; aramid fiber; ZrB_2; high silica fiber scrim. Base rubber and various fillers were pre-mixed and infiltrated for more than four hours, then dispersed by high speed dispersion for half an hour. After that TEOS with a mass fraction of 3% and the mass fraction of 0.5% organic tin were mixed into the mixture, the material was stirring evenly and pressed into piece by a mould after discharging air bubbles.

According to the actual experience system demand, the finished product was cut into specimen with size of 50×50×10 mm, whose density is about 1.58g/cm^3.

2.2. Ablation test system and analytical instruments

The ablation test system used in materials property testing uses aviation kerosene and oxygen gas as combustion agents and oxidizers. The particle concentration of two-phase flow is controllable (0~30%), and the ablation angle can be adjusted between 0~90°, Fig. 1 shows the schematic diagram of the ablation experiment system.

2.3. Ablation test parameters

In the experiment, the powder injecting to the flame is boron oxide whose size is 15~45μm, and the purity is higher than 99.2%. Fig. 2 is the micro morphology of boron oxide powder, the powder is round in shape and has a uniform particle size.

The analytical instruments used in the experiment include the scanning electron microscope produced by the Czech TESCAN company and the Energy350 INCA energy dispersion spectrum provided by the British OXFORD company.

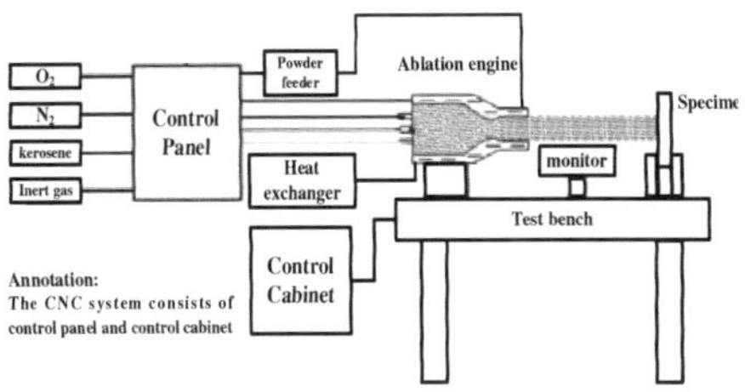

Figure 1. Schematic diagram of the experiment system

Figure 2. Morphology of the boron oxide powder (×300)

2.4. Ablation test parameters

The oxygen enrichment of the test condition has a great influence on the ablation result, and the value of β is determined by the following formula (1). k_0 means the theoretical mixture ratio of kerosene and oxygen. According to the kerosene brand, the value is 3.3.

$$\beta = \frac{m_1 - k_0 m_2}{m_1 + m_2 + m_3} \quad (1)$$

The particle concentration η of oxygen-kerosene simulation ablation test system is the mass fraction of the particle in the whole flame jet, the calculation formula of η is shown in (2):

$$\eta = \frac{G}{Q_1 \times \rho_1 + Q_2 + Q_3 \times \rho_3 + G} \quad (2)$$

The specific parameter setting is shown in Table 1. Under this condition, the temperature of the jet touching surface of the specimen was about 2000K, the speed was about 470m/s by thermal calculation and simulation analysis, which could meet the practical requirements of the thermal insulation material ablation of the solid ramjet engine. A total of two groups of ablation conditions were designed in the test, with or without particles erosion as contrast variables. Three specimens were set of each group under the ablation condition.

Table 1. Ablation test scheme of system for thermal insulation materials

Specimens	A1 A2 A3	B1 B2 B3	Specimens	A1 A2 A3	B1 B2 B3
Oxygen flux/(L·min-1)	342	342	Ablation	80	80
Kerosene flux/(kg·min-1)	0.138	0.138	Ablation angle/°	45	45
Oxygen-rich condition/%	5	5	Combustion	0.27	0.27
Carry gas flux/(kg·min-1)	20	20	Test time/s	20	20
Particle concentration/%	0	1.69			

3. Results and discussion

3.1. Ablation rate measurement

The ablation rates of the specimens are calculated, and the final ablation rates are shown in Table 2.

Table 2. Average ablation rates of the specimens

No.	Mass ablation rate/($g \cdot s^{-1}$)	Linear ablation rate/($mm \cdot s^{-1}$)	No.	Mass ablation rate/($g \cdot s^{-1}$)	Linear ablation rate/($mm \cdot s^{-1}$)
A1 A2 A3	0.0099	-0.025	B1 B2 B3	0.1775	0.437

Under the condition of pure gas ablation, the linear ablation rate of the material is negative. With the addition of particles, the two kinds of ablation rates of the specimens have been greatly increased, which shows that the erosion of particles is the main reason for the insulation material consumption.

3.2. Macroscopic ablation morphology analysis

After ablation, the three specimens in the A group overall expanded, whose flame contact surface became rough and turned black. No ablation pit appeared in the center of the specimens in the A group, the macroscopic morphology of A2 specimen is obtained as shown in Figure 3(a) after it was cut. The section morphology shows that the specimen formed a layered morphology after the ablation of the insulating materials, and the reaction layer (ceramic layer and pyrolysis layer) at the center of the ablation is slightly thicker than the edge, the naked eye can directly see the interface between the reaction layer and the virgin layer, but the ceramic layer and the pyrolytic layer in the reaction layer were not easy to distinguish.

After ablation, the ablation pits were observed in the center of the group with particle, Figure 3(b) shows B3's overall appearance (including specimen holder). The ablation pit is circled out shown as the A area in this figure. There were bundle ablation gully morphology below the ablation pit of specimens along 45° direction that was the ablation jet spray direction, because of the high temperature, high speed and high heat flux of the jet center, the damage to the center of the material was stronger than the jet edge, and finally the pit was left on the surface of the material; Partial white crystal was retained in the specimen B area, and the upper half of the specimen and the C area of the specimen holder were covered with dark brown substance.

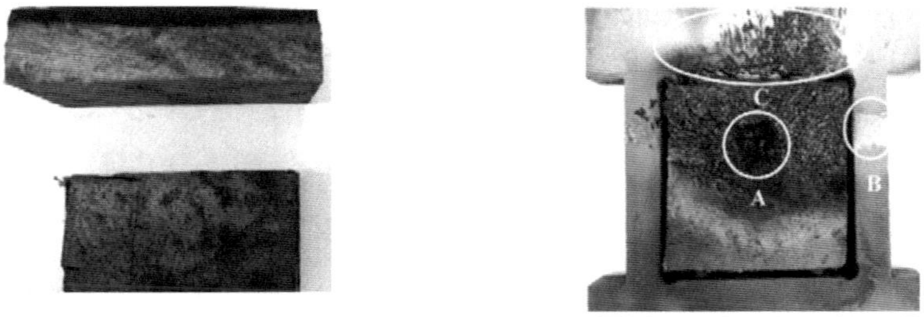

(a) Without erosion (specimen A2) (b) With erosion (specimen B3)

Figure 3. Ablation morphology of the specimens

3.3. Microscopic morphology analysis

In order to investigate the ablation mechanism of silicone rubber based insulation material under particles erosion of boron oxide condition, the micro morphology and material composition of the material were analyzed by SEM and EDS.

3.3.1 Microscopic morphology of specimens after ablation under pure gas ablation

Fig. 4 is the morphology of the ceramic layer and the pyrolysis layer of the specimen A3 at different magnification. The phenomenon of delamination in the ablation region can be clearly observed in Fig. 4(a). Figure (b) and (c) in the Fig. 4 respectively are the local amplification morphology of the ceramic layer and the pyrolysis layer. The structural features of the ceramic layer and the pyrolysis layer are obvious, and the ceramic layer structure is much compact， which can play a better role in protecting the parts under ceramic layer, at the same time, heat passes through the ceramic layer and passes inwards, Pyrolysis of internal polymeric materials brings the formation of a large amount of gas, which results in larger internal pressure in the material. With the overflow of gas, a unique hole in the pyrolysis layer is formed. The area A and B of Figure 4(c) and whole area shown in 4(b) are analyzed for atomic percent by EDS, and the results obtained as Table 3 shown, as we can see, the main elements of ceramic layer are carbon, silicon and oxygen, it is inferred that the main component is fused silicon dioxide with a small amount of silicon carbide which has not been oxidized and pyrolysis by flame gas, they are densely covered with particles on the specimen surface. The material in pyrolysis layer is mainly silicon dioxide, some of which are as the wall surface of pores in the form of molten state. Some of them are attached to the wall of holes in the form of particles, with the release of a large amount of gas from pyrolysis process of polymer materials, material expand and the expansion process cannot be effectively restrained, so the final linear ablation rate is negative value. The ceramic layer effectively blocks the erosion of the pure gas jet, and the pyrolysis process takes away a lot of heat. These two processes ensure the reliability of the insulation material.

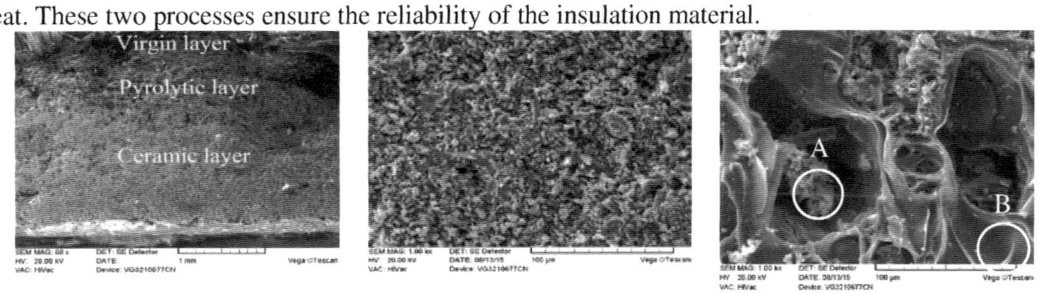

(a) Section (×60) (b) Ceramic layer (×1000) (c) Pyrolytic layer (×1000)

Figure 4. Micro-morphology of different structural layers

Table 3. The atomic percent of the ceramic layer and pyrolytic layer

Element	Ceramic layer	Pyrolytic layer(A)	Pyrolytic layer(B)
C	6.17	8.51	5.13
O	51.95	66.44	74.37
Si	37.88	24.61	20.50

3.3.2 Microscopic morphology of specimens after ablation under boron oxide erosion condition

The addition of boron oxide particles causes a large change to the ablation morphology. At the center of specimen morphology is shown in Figure 5(a), the specimen surface attached with a laminar like molten material, which illustrates that the flame temperature reaches the melting point of material under the test conditions, after further amplifying white frame area in the figure 5(a), we can get the morphology shown in figure 5(b), it is observed that molten material have a certain degree of rupture and have not completely covered on the material surface with partial ceramic layer areas exposed to flame and touching to the particles when ablated; at the same time, some large volume particle are not fused completely, and embedded in the molten layer by solid state by the driving of high velocity flame stream, as shown by the white particles on the surface in the figure 5(b). Table 4 is the results of elemental analysis, which shows that the surface white particles are mainly oxygen and boron elements, it can be preliminarily judged that the white particles are boron oxide powder.

The morphology of 45° jet scouring area near the ablation pit is shown in Figure 6. Compared with the central area, the coverage of the melt in the scouring area for the material is more complete, and the cracks are more small. The reason is that the central area is subjected to the positive impact of particles and its force environment is worse than scouring area. But in the scouring area, particles fly following the arrow shown in the fig.6, leading to the gully morphology.

Section morphology of the specimen B3 is show as figure 7. Compared with the section morphology of the specimen A2 that was ablated under pure gas condition, the morphology shown in the fig. 7 has a thinner reaction layer that can still be observed clearly, which shows that the addition of solid particles in the gas has a strong erosion and destruction to the surface ceramic layer. Besides, a layer of molten boron oxide is also deposited over the reaction layer to reduce permeation of the oxidizing components and slow down the rate of oxidation depletion of materials that is below surface. However particle erosion is the main cause of material depletion, the combined effect of the above both led to the upward material ablation rates of silicone rubber.

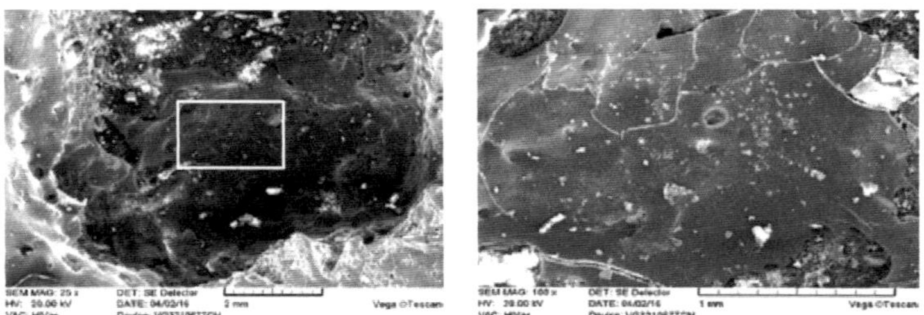

(a) (×25) (b) (×100)

Figure 5. Morphology in the central area of specimen

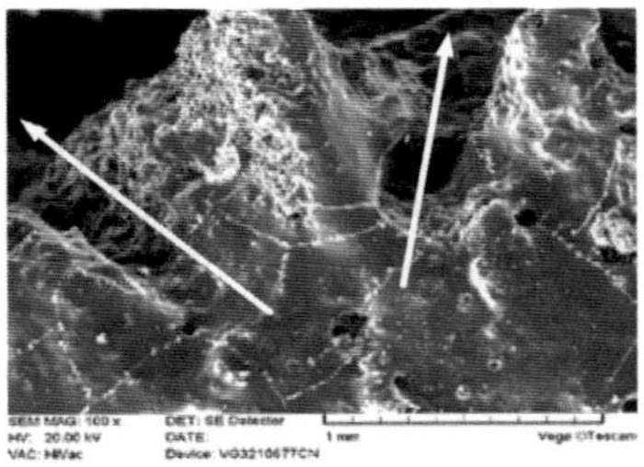

Figure 6. Morphology in the scour area (×100)

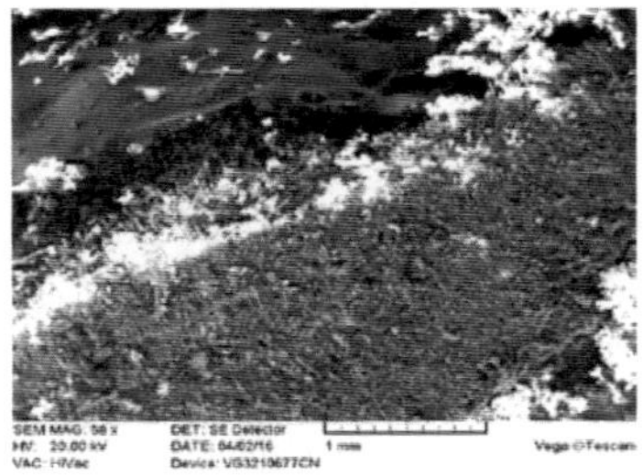

Figure 7. Section (×58)

Table 4. The atomic percent of the central area of specimen

Element	Atomic percent	Element	Atomic percent
B	36.14	O	44.87
C	15.39	Si	3.60

Microscopic of the area B and C in Fig. 3 are shown in fig. 8. The white crystalline in the area B is boron oxide according to the EDS analysis results shown in Table 5. After recrystallization, powders condensed into balls and pile together with small tiny bumps attaching to the balls surface. The overall size of the balls is slightly less than boron oxide powder of addition, the shape of the particles changed from irregular to spherical, this process of change shows that the particles are fully molten in the flame gas, which indirectly proofing the flame under this experimental condition has a similar thermal environment to the secondary combustion chamber of solid rocket ramjet engine using boron propellant; And EDS result of the area C show that the main elements in this area is boron, oxygen and carbon elements, referring to the B area result, the dark brown material in the area C includes not only the molten boron oxide in the gas stream but also the small amounts of silicon dioxide and unoxidized silicon carbide that migrated from the surface of the specimen.

Table 5. The atomic percentage of the element in the area B and area C in specimen handle

Element	Area B	Area C	Element	Area B	Area C
O	59.55	39.26	C	—	21.60
B	40.45	33.10	Si	—	6.03

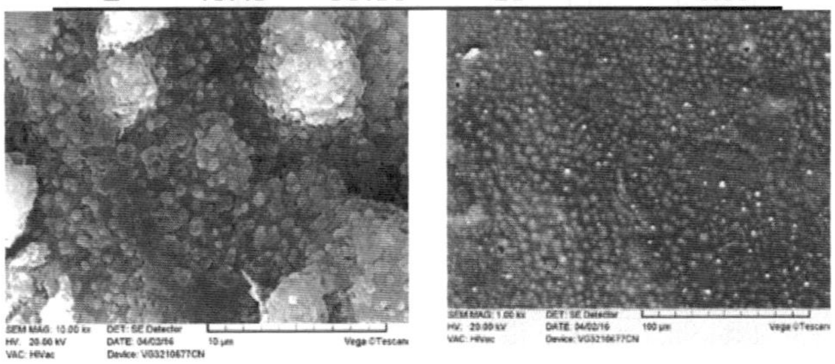

(a) Area B (×10000) (b) Area C (×1000)

Figure 8. Microscopic morphology of corner ceramic coatings under different oxygen – enriched conditions

4. Conclusion

(1) In this paper, the ablation test of silicone rubber based insulation materials under the condition of Boron Oxide particle erosion is carried out. After calculation, the average mass ablation rate of the material under the multiphase flow test is 0.1775 $g \cdot s^{-1}$, the average linear ablation rate is 0.437$mm \cdot s^{-1}$. In the pure gas phase contrast test, the linear ablation rate is negative value, and the addition of erosion particles is the main factor affecting the depletion of the material.

(2) Boron oxide particles after heating and accelerating attached to the surface of the material and formed a layer of molten material, which covering material pores leading to blocking the infiltration passage of the oxidizing component and slowing down the rate of oxidation depletion of materials below surface. However particle erosion is the main cause of material depletion, the combined effect of the above both led to the upward material ablation rates of silicone rubber.

(3) The boron oxide particles doped in the high temperature flame flow of oxygen kerosene have reached the molten state, which is close to the actual combustion flame flow of solid rocket ramjet engine using boron propellant. The simulated ablation method is better for the reconstruction of the ablation environment. The simulated ablation method used in this article has high application value in the reconstruction of ablation environment.

References

[1] Iqbal N, Khan M B, Sagar S, et al. Fabrication and characterization of multiwalled carbon nanotubes/silicone rubber composites[J]. Journal of Applied Polymer Science, 2013, 128(4):2439-2446

[2] Liu Y, Guan Y W, Wu Y F, et al. Experimental Investigation on Ablation Characteristic of Silicone Rubber Insulator under Particle Erosion Condition[J]. Journal of Propulsion Technology, 2017, 38(1): 227-234

[3] Standard Test Method for Oxyacetylene Ablation Testing of Thermal Insulation Materials[S]. ASTM E285-2008

[4] CHEN C J, MA G F. Study on Internal Insulation of Rocket Engine in Oxidizer Rich Environment[J]. Journal of Rocket Propulsion, 2004, 30(4): 32-35

[5] WANG S X, HE G Q, LI J, et al. Experimental Investigation on Ablation Characteristics of Silicon Rubber Insulator in Different Gas Environments[J]. Journal of Solid Rocket Technology, 2009, 32(5): 583-587

[6] Zhang Y, Chen G H, Wang J G. Study on Thermal Protection of Silicon Rubber Insulation Used in the Ramjet Afterburner[J]. Chinese Journal of Explosives and Propellants, 2007, 30(3): 65-68

[7] Zha B L, Huang D Y, Qiao S L, et al. Research on ablation test and ablative mechanism of carbon / carbon composite material[J]. Journal of Solid Rocket Technology, 2013, 36(5): 692-696

[8] Li L, Yang T, Cheng X H, et al. Ablation Model of Silicon Insulator in Ramjet Combustion Chamber[J]. Journal of Propulsion Technology, 2012, 33(3): 451-454

[9] Yang D. Silicone Rubber Based Insulation Material and Its Thermochemical Ablation Mechanism[D]. Changsha: National University of Defense Technology, 2013

[10] Li L, Yang T, Jiang W. Effcets of High-velocity Particles on Solid Ramjet Insulator[J]. Journal of Projectiles Rockets Missiles and Guidance, 2012, 32(1): 127-130

[11] Wang S X, Li J. Experimental Analysis of Silicon Rubber Insulator Thermo-chemical Ablation Characteristics[J]. Science Technology and Engineering, 2015, 15(15): 218-221

[12] Yang D, Zhang W, Jiang B. Ceramization and oxidation behaviors of silicone rubber ablative composite under oxyacetylene flame[J]. Ceramics International, 2013, 39(2):1575-1581

IOP Conf. Series: Materials Science and Engineering **303** (2017) 012004 doi:10.1088/1757-899X/303/1/012004

Relationship between compatibilizer and yield strength of PLA/PP Blend

Pattanun Jariyakulsith[1] and Somchai Puajindanetr[2]
[1,2] Department of Industrial Engineering, Faculty of Engineering, Chulalongkorn University, Bangkok, 10330, Thailand

[1]ploy_pattanun@hotmail.com, [2]fiespj@hotmail.com

Abstract. The aim of this research is to study the relationship between compatibilizer and yield strength of polylactic acid (PLA) and polypropylene (PP) blend. The PLA is blended with PP (PLA/PP) at the ratios of 70/30, 50/50 and 30/70. In addition, (1) polypropylene grafted maleic anhydride (PP-g-MAH) as a compatibilizer at 0.3 and 0.7 part per hundred of PLA/PP resin (phr) and (2) dicumyl peroxide (DCP) being an initiator at 0.03 and 0.07 phr are added in each composition. Yield strength is characterized to study the interaction between compatibilizer, initiator and yield strength by using experimental design of multilevel full factorial. The results show that (1) the yield strength of PLA/PP blend are increased after addition of compatibilizer. Because the adding of PP-g-MAH and DCP resulted in improving compatibility between PLA and PP. (2) there are interaction between PP-g-MAH and DCP that have affected the final properties of PLA/PP blend. The highest yield strength of 27.68 MPa is provided at the ratio of 70/30 blend by using the 0.3 phr of PP-g-MAH and 0.03 phr of DCP. Linear regression model is fitted and follow the assumptions of normal distribution.

1. Introduction

The number of plastic waste increases in recent years. To concern the environmental impact, the biodegradable materials are play in a crucial role in daily life. Thus, this research interested in biodegradable polymer called polylactic acid (PLA) which is derived from the starch or sugar cane [1,2]. PLA widely used especially in disposable packaging materials such as bags, food packaging, home appliances and compostable tableware, because of its good mechanical properties [2-4]. However, the limitations of PLA are rigid and brittle at room temperature [2,4], so its applicability has been limited. PLA might be improved by using the techniques such as melt blending with tough polymers, plasticization, and chemical modification [3]. Hence, PLA which was blended with soft and tough polymers can lead to enhance mechanical properties and biodegradability [4]. Polypropylene (PP) is a commodity polymer with worldwide production, low cost, easiness of processing and non-toxicity [5]. However, the properties of polymer blend are limited because of the limited compatibility between phases of polymers. To solve this problem by enhancing the adhesion between phases, the compatibilizers are often used as additives [6]. Polypropylene grafted maleic anhydride (PP-g-MAH) is well known as an effective compatibilizer for blending PP with polar polymers (i.e. PLA). Maleic anhydride is a famous initiator because of chemically good react with the PLA, availability, low toxicity and low potential to react in itself during the process [7]. Some paper was studied about effect of PP-g-MAH as a compatibilizer on PLA/PP blend, the morphology analysis of PLA/PP blend confirmed that filling of PP-g-MAH can lead to adhesion between PLA and PP. From the Scanning Electron Microscopy of PLA with PP-g-MAH, showed the ductile fracture due to interaction of PP-g-MAH and

PLA [5]. For more details, the study of mechanism of melt grafting observed that the amount of maleic anhydride correlated with the level of DCP which effect to the properties [8]. Another research used the PP-g-MAH to improve the compatibility of PLA/PP blend, confirmed that PP-g-MAH was coupled with PP by the PP part in its molecule and the anhydride part reacted with the polar part in PLA molecule. Furthermore, it can be seen that the PLA improved the biodegradability of PP [9]. So the PLA/PP blend would become a new alternative to be a biodegradable materials. To define all the factors that effect on desirable properties in order to reduce cost, production time, even the optimal condition of processing, the experimental design should be involved. This research aimed to improve the mechanical properties of PLA by blended with PP in order to approach the new material that degradable, good mechanical properties and the lowering cost simultaneously. To study the relationship between the PP-g-MAH as a compatibilizer and DCP as an initiator to the properties of PLA/PP blend which under the ratios of 70/30, 50/50 and 30/70 by using the experimental design method. In addition, (1) PP-g-MAH at 0.3 and 0.7 phr and (2) DCP at 0.03 and 0.07 phr were added in each composition. Yield strength are investigated. For the future, PLA could formed by various polymer processing and could have more applications without the limitations.

2. Materials and methods

2.1. Materials
Polylactic acid (PLA) resin, 4032D grade (available in pellet form) was purchased from NatureWorks LLC. Polypropylene resin, HP6060 grade were supplied by HMC polymers. Polypropylene grafted maleic anhydride (PP-g-MAH) was purchased from Sigma-Aldrich as a compatibilizer. Dicumyl peroxide (DCP), supplied by Kit Phaibun Chemistry Ltd, was used as an initiator.

2.2. Preparation of blending of PLA/PP
The Blending of PLA/PP using the extrusion via twin screw extruder, with/without initiator and compatibilizer. The conditions for twin screw extruder were as follows: (1) Screw speed at 40 rpm,
(2) Temperature of 10 zones were (175/180/190/190/195/195/195/195/180/180°C respectively).
In addition, the pellets of polymers were dried all night at 55 °C before processing. The PLA/PP blend was carried out by extruder, then compressed to be the specimen by compression moulding at 200 °C, pre heating for 20 minutes, and pressure of 1.96 MPa for 5 minutes.

2.3. Testing and characterization
Yield strength were tested according to ISO527-2 by tensile testing machine as a model of 112.100kN.H. from MMT engineering co., ltd. The conditions of testing were as follows: (1) cross head speed of 5 mm/min and (2) load cell of 100kN.

2.4. Experimental design
As mentioned above, the proper amount of compatibilizer is an important point to consider, because it would effect on the final properties of polymer blend. Therefore, the experimental design would play a role which can reach about interaction between factors and responses with precise estimation in main effects of each factor and their interactions. In this study selected multilevel full factorial design using the Minitab17 with random of five replicates and the levels of factors are shown in table 1.

Table 1. Factors and levels for experimental design

Factors	Levels		
	Low	Medium	High
Factor A : PLA (%)	30	50	70
Factor B : PP-g-MAH (phr)[a]	0.3	-	0.7
Factor C : DCP (phr)[a]	0.03	-	0.07

[a] part per hundred of PLA/PP resin

3. Results and discussions

3.1. Analysis of variance

To understand the significant of factor to yield strength, the analysis of variance from Minitab 17 was used which shown in figure 1.

Source	DF	Adj SS	Adj MS	F-Value	P-Value
Regression	9	748.427	83.1585	28.53	0.000
PLA	1	14.600	14.5995	5.01	0.030
PP-g-MAH	1	0.153	0.1525	0.05	0.820
DCP	1	19.316	19.3163	6.63	0.013
PLA*PLA	1	55.972	55.9716	19.20	0.000
PLA*PP-g-MAH	1	2.889	2.8888	0.99	0.324
PLA*DCP	1	64.579	64.5790	22.15	0.000
PP-g-MAH*DCP	1	26.215	26.2149	8.99	0.004
PLA*PLA*PP-g-MAH	1	20.783	20.7834	7.13	0.010
PLA*PP-g-MAH*DCP	1	76.121	76.1208	26.11	0.000
Error	50	145.747	2.9149		
Lack-of-Fit	2	3.273	1.6365	0.55	0.580
Pure Error	48	142.474	2.9682		
Total	59	894.174			

Figure 1. Analysis of variance

From the figure 1, can conclude that there are two types of significant factor (p-value < 0.05) which consist of (1) main effects; the content of PLA and content of DCP. And (2) interaction effects that are interaction between PLA/DCP and interaction between PP-g-MAH/DCP. All of these factors are significantly affect to the yield strength of PLA/PP blend.

3.2. Main & interaction effects

To understand the relationship between compatibilizer and yield strength, the Minitab 17 is needed. Because some factor is not directly affect the yield strength but it might influence under interaction with another factor. So in this research will describe by the main effect plot and interaction plot which are shown in figure 2 to easily understand.

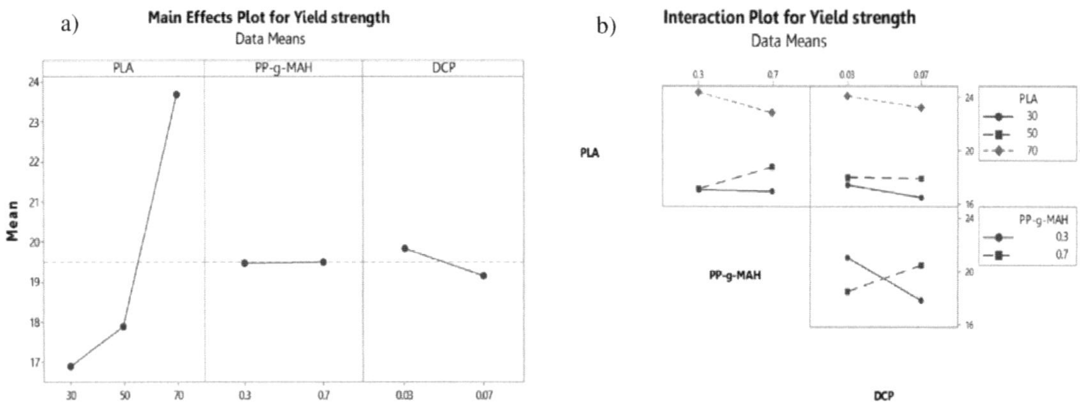

Figure 2. (a) main effects plot, (b) interaction effects plot for yield strength.

As the figure 2(a) and 2(b) show main effects plot and interaction effects plot for yield strength respectively. In main effects plot describe that the yield strength are changed when the content of PLA increases and show the enormous changing from medium to high level of PLA. The addition of only PP-g-MAH seems not to affect the yield strength, it need to be used with another factor to notice the different. In the contrary, the increasing of DCP responsible for the yield strength to decrease. For the interactions plot, shows the interaction between PLA and PP-g-MAH are as follows: (1) at the high level of PLA, if add the PP-g-MAH can lead to the reducing of yield strength while (2) addition of PP-g-MAH at low level and medium level of PLA causes gradually increase in yield strength. Next, the interaction between PLA and DCP which effects to yield strength can describe as follows: (1) at high level and low level of PLA, increasing in DCP can lightly effect on yield strength in negative way. Conversely, (2) filling of DCP bring about the yield strength to be higher at the medium level of PLA. And the last is interaction between PP-g-MAH and DCP are reversely effect to the yield strength, it means that changing from low level to high level of DCP at the low level of PP-g-MAH provides the yield strength to be poor, while at the high amount of PP-g-MAH with increasing in DCP gives the raising of yield strength.

3.3. yield strength in each composition

In the figure 3 as below, it can be seen that, the greater amount of DCP at 0.3 phr of PP-g-MAH in each composition can result in the lower yield strength. It may be concluded that the excessive amount of DCP can lead to the overwhelming of free radicals in blending, then induced the chain of polymer to entangle as a result of diminished yield strength. On the contrary, the changing of DCP from low to high level at constant 0.7 phr of PP-g-MAH can lead to the higher yield strength. Because at the more amount of PP-g-MAH which means there is some remained of PP-g-MAH in reaction, so the free radicals from DCP can react with PP-g-MAH, PLA and PP. In addition, each composition of PLA/PP blend can summarize into two cases as follow; (1) at the ratio 30/70 and 70/30 of PLA/PP, using the 0.3 phr of PP-g-MAH and 0.03 phr of DCP will provide the highest yield strength but (2) at ratio of 50/50 of PLA/PP, using 0.7 phr of PP-g-MAH and 0.07 phr of DCP gives the highest yield strength. It may conclude that at the different ratio of polymer, the optimal content of used compatibilizer and initiator would not be the same amount. Thus, approaching the optimal solution by using the experimental design will be useful in the future.

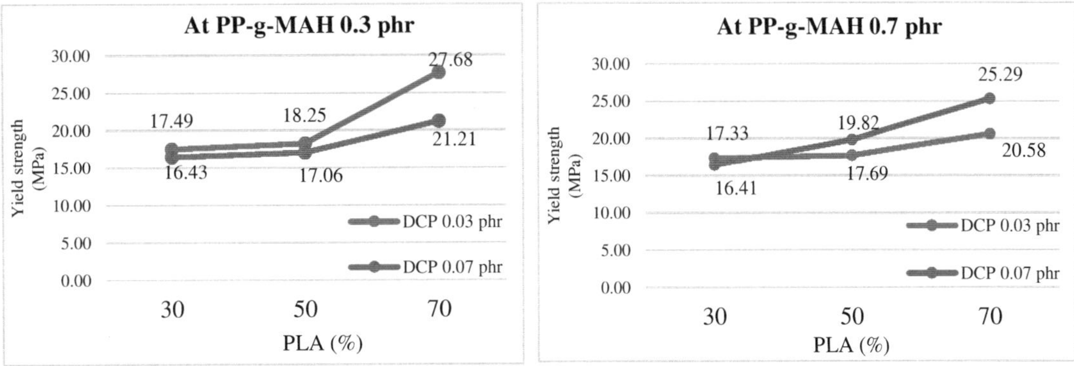

(a) At PP-g-MAH 0.3 phr (b) At PP-g-MAH 0.7 phr

Figure 3. Yield strength of PLA/PP blend in each composition with different amount of DCP

3.4. Regression model

For develop the effective model, the normal probability plots of residual for yield strength follows the assumption of normal distribution and the variances are at constant. In this study, R^2 (%) and regression model of yield strength is shown in table 2. It reveals the R^2 are closer to 100%, so this model is an effective tool to predict the mechanical properties as a requirement in commercial industry.

Table 2. Regression reduced model for response.

Response	Regression reduced model*	R^2 (%)
Yield strength (MPa)	Yield strength = 27.27 - 0.738A - 3.7B + 246.0C + 0.01379 A^2 + 0.609AB - 8.55AC – 532BC - 0.01561 A^2B + 17.24ABC	83.70

*A,B,C are the factors in table1

From the regression model, can conclude that the decreased of one phr in PLA leads to reduce the 0.738 MPa of yield strength and also in PP-g-MAH that decreased of one phr responsible for declined 3.7 MPa of yield strength. In the other hand, the adding of one phr of DCP results in the higher of yield strength to 246 MPa. Not only the main effect but also the interaction effect that influence to the yield strength as shown in the model.

4. Conclusion

The investigation of this study provide the regression model that fitted the data and resulting in the effective forecast to mechanical properties which can practical in production with the required properties. From this study, the improvement of yield strength depends on the amount of compatibilizer, initiator and interaction between them. From overall of the results, the yield strength of PLA/PP blend are improved after adding the compatibilizer which result from enhancing the adhesion between phases as shown in figure 4. PLA/PP blend (70/30) with 0.3 phr PP-g-MAH and DCP 0.03 phr has the highest yield strength of 27.68 MPa. However, the tough and soft polymer may act as a plasticizer in the blending of polymer, so addition of plasticizer more than 20-30% into PLA matrix can cause the phase between polymer blend to separate [2]. Hence, the changing of properties could not clearly distinguishable.

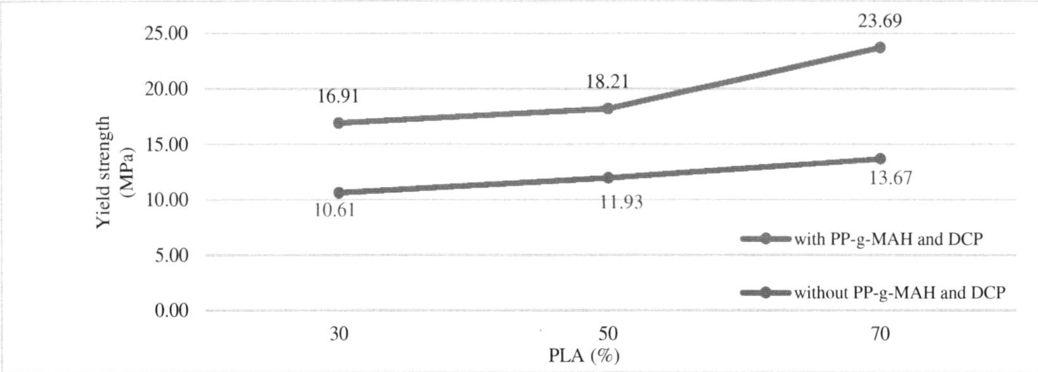

Figure 4. Yield strength of PLA/PP blend with and without PP-g-MAH and DCP

References

[1] Detyothin, S., Selke, S.E.M., Narayan, R., Rubino, M. and Aurus, R. (2013) Reactive functionalization of poly(lactic acid), PLA: Effects of the reactive modifier, initiator and processing conditions on the final grafted maleic anhydride content and molecular weight of PLA. *Polymer Degradation and Stability*, **98**(12)**,** 2697-2708.

[2] Hassouna, F., Raquez, J.M., Addiego, F., Dubois, P., Toniazzo, V. and Ruch, D. (2011) New approach on the development of plasticized polylactide (PLA): Grafting of poly(ethylene glycol) (PEG) via reactive extrusion. European Polymer Journal, 47, 2134-2144.Wit, E. and McClure, J. (2004) Statistics for Microarrays: Design, Analysis, and Inference. 5th Edition, John Wiley & Sons Ltd., Chichester.

[3] Ku Marsilla, K.I. and Verbeek, C.J.R. (2015) Modification of poly(lactic acid) using itaconic anhydride by reactive extrusion. *European Polymer Journal*, **67**, 213-223.Giambastiani, B.M.S. (2007) Evoluzione Idrologica ed Idrogeologica Della Pineta di san Vitale (Ravenna). Ph.D. Thesis, Bologna University, Bologna.

[4] Hamd, K., Kaseem, M. and Deri, F. (2012) Poly(lactic acid)/low density polyethylene polymer blends: preparation and characterization. ASIA-PACIFIC JOURNAL OF CHEMICAL ENGINEERING, 7, 310-316.

[5] Pivsa-Art, S., Kord-Sa-Ard, J., Pivsa-Art, W., Wongpajan, R., O-Charoen, N., Pavasupree, S. and Hamada, H. (2016) Effect of Compatibilizer on PLA/PP Blend for Injection Molding. *Energy Procedia*, **89,** 353-360.

[6] Ebadi-Dehaghani, H., Khonakdar, H.A., Barikani, M. and Jafari, S.H. (2015) Experimental and theoretical analyses of mechanical properties of PP/PLA/clay nanocomposites. Composites Part B: Engineering, 69, 133-144.

[7] Hwang, S.W., Lee, S.B., Lee, C.K., Lee, J.Y., Shim, J.K., Selke, S.E.M., Soto-Valdez, H., Matuana, L., Rubino, M. and Auras, R. (2012) Grafting of maleic anhydride on poly(L-lactic acid). Effects on physical and mechanical properties. *Polymer Testing*, **31**(2), 333-344.

[8] Shi, D., Yang, J., Yao, Z., Wang, Y., Huang, H., Jing, W., Yin, J. and Costa, G. (2001) Functionalization of isotactic polypropylene with maleic anhydride by reactive extrusion: mechanism of melt grafting. *Polymer*, **42**(13), 5549-5557.

[9] Ploypetchara, N., Suppakul, P., Atong, D. and Pechyen, C. (2014) Blend of Polypropylene/Poly(lactic acid) for Medical Packaging Application: Physicochemical, Thermal, Mechanical, and Barrier Properties. *Energy Procedia*, **56**, 201-210

IOP Conf. Series: Materials Science and Engineering **303** (2017) 012005 doi:10.1088/1757-899X/303/1/012005

Analysis of the effects of stirring condition of separation of thorium in the elution process of monazite partial solution by solvent impregnated resin method

R Prassanti[1], D S Putra[2], B P Kusuma[2] and F W Nawawi[2]

[1]Center for Nuclear Minerals Technology, National Nuclear Energy Agency of Indonesia (BATAN)
[2]Metallurgical and Materials Engineering, Universitas Indonesia, Kampus UI Depok, 16424, Jawa Barat, Indonesia

dwiki.syahbana@ui.ac.id

Abstract. Monazite is a natural mineral which contains abundant valuable element such as Radioactive Element and Rare Earth Element(REE). In this experiment, it is proven that solution of residual Thorium Sulfate from Monazite mineral process, can be seperated selectively by using extracting method of Solvent Impregnated Resin(SIR), with the elutant solution HNO_3. In the earlier process, Thorium solution is conditioned at PH 1 by using H_2SO_4. Then REE, Thorium and Uranium elements are seperated. This seperation is conducted by using adsorption method by Amberlite XAD-16 Resin, which has been impregnated by Tributhyl Phosphate extractant. It is continued with elution process, which is aimed to obtain Thorium solution of a higher level of concentration. This elution process is conducted by using HNO_3, with the elution variables of the lenght of mixing and amount concentration elutant. Based on this experiment, SIR extracting method is able to dissolve Thorium solution until 63,2%grade and a higher level of %grade about 92,40%. It can be concluded that this SIR method can extracted Thorium elements selectively, improve extracting process recovery, and determine optimum stripping condition in the 45th minutes with elutant concentration of 1,0M HNO_3.

1. Introduction

Mineral Monazite contains valuable elements that can be further processed such as U, Th, and RE. In order to be processed, high purification of Thorium is required as in the use of nuclear fuel, as much as > 87.42%[1]. The extraction process is aimed on obtaining high concentration and pure elements. In the previous research, there has been a test of increasing the amount of Thorium extracts by using ion exchange resin method. However, Thorium cannot be separated with Uranium. The stripping method is a proper separation step between Thorium with another elements. Stripping Thorium can be selectively taken from Tributyl Phosphate (TBP) extraction using acid solution, whereas Uranium can be taken by using demineralized water[2]. The efficiency of stripping using TBP with 5M HNO3 solution condition was 96.1%[3]. TPB is a non-polar organic extractant with good sorption ability against Amberlite XAD-16 Resin[4]. So then stripping extraction is optimized by ion exchange mechanism by Amberlite XAD-16 Resin.

2. Materials and Method

2.1 Feed Preparation
The Partial solution feed is prepared as a solution that is obtained from the previous process of Monazite in the form of partial solution of Th-U-RE. Partial solution is then extracted to get high recovery of thorium element using SIR method. The Thorium solution is conditioned on the acid condition of PH 1 by using H2SO4 solution. The feed is prepared as much as 3000ml, with a thorium content of 28480.0mg. Amount of thorium elements is obtained from the characterization of ICP-OES.

2.2 Resin Amberlite XAD-16 Preparation
In the SIR extraction method, it is necessary to prepare the resin as an absorbent element of Thorium in the feed. The resin in this method is Amberlite XAD-16 Resin. This resin is used in the Solid Phase Extraction, because of their physical and chemical features such as high surface areas, porosity, purity, and durability[5]. Resin preparation is done through the impregnation process of Tributyl Phosphate (TBP) extractant. Prior to the impregnation process, Amberlite resin is cleaned by using 2M HNO3, NaOH 2M and acetone. Then it is dried at a temperature of 50°C in the oven.

2.3 Elutant Preparation
A strong acid solute eluent is used for the elussion process. Elutant is a solvent for separating the thorium elements selectively from Amberlite resins. The elutant used in this experiment is HNO3 with various concentrations. The HNO3 solution concentration is obtained from dilution using concentration test on 0.5M, 1.0M, and 1.5M and aquades(0M).

2.4 Tributyl Phosphate
Tributyl Phosphate (TBP) is an extractant solution which has been proven to be able to extract Thorium element. The TBP is a non-polar organic solution that can bind with Amberlite resin well. The molecular size and polarity of TBP are known capable to produce adsorption capacity of Amberlite XAD-16 untill 900mg/g [4]. TBP is mixed with acetone to reduce the viscosity of the solution also the surface areas, pore volumes, and pore sizes of Amberlite resin[10]. It is therefore expected to be better adsorbed in the Amberlite resin

2.5 Adsorption Process
Thorium feed in this experiment is obtained through the adsorption step by Amberlite XAD-16 Resin. There are 30g impregnated Amberlite resin immersed in a 3000ml partial solution with a Thorium content of 28480.0mg. The impregnation process is performed on 3000ml Thorium feed and 30ml of TBP extractant. The process is terminated by drying the Amberlite resin in the oven at 50 °C. The adsorption process produces a new initial feed content in the SIR extraction method..

2.6 Elution Process
The result of impregnation on Amberlite resin is separated by HNO3 elution process. The elution process is carried out under conditions of HNO3 concentration as well as in different stirring times. Elution is performed at a concentration of 0.5M, 1.0M, 1.5M HNO3 with the stirring duration (5, 15, 30, 45, 50) minutes.

3. Result and Discussion

3.1 Mechanism of Amberlite XAD-16 Resin as an Ion Exchange Media
The elution process uses an Amberlite-XAD16 resin that has been impregnated by an organic solution

of TBP. TBP in this case is known as an extracting solution in Thorium. It has a chemical structure between the TBP and the Amberlite XAD-16 resin as shown in figures 1 and 2.

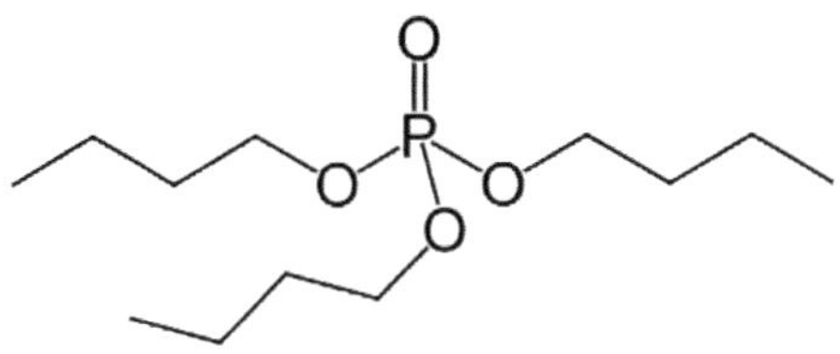

Figure 1. Tributyl Phosphate Structure

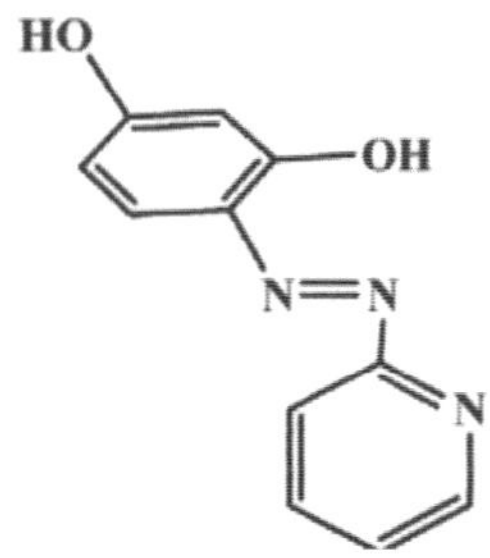

Figure 2. Amberlite XAD-16 Structure[9]

The impregnation process occurs complexly between ligand and Thorium. The ligands will be loaded by Amberlite XAD-16 Resin from TBP. The ligands present in the solid phase will react with Th4+, forming the neutral complex Th (nBPHA) 4, which is present in the XAD process [6]. Interaction can be seen in figure 3.

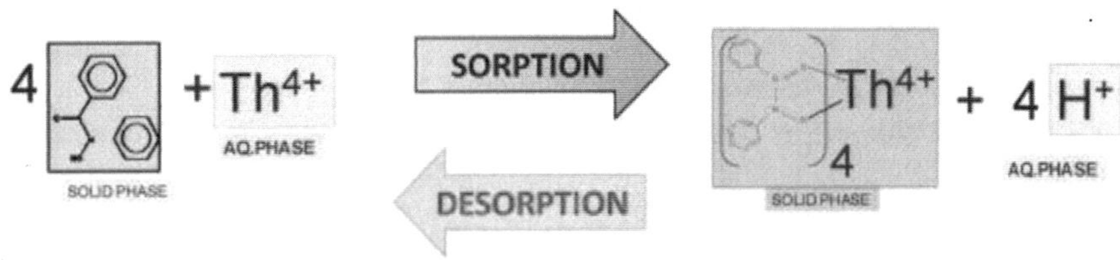

Figure 3. Interaction Between Solid phase and Thorium

So the equation is generated as in the model in the following equation, (S.Michaud. Modelling of Thorium Extraction by TBP)

Extraction equilibrium of Th(IV) bv TBP with nitrate ion is:

$$Th_{aq}^{4+} + 4NO_{3aq}^{-} + nTBP_{org} \leftrightarrow (Th(NO_3)_4, nTBP)_{org} \qquad (1)$$

Or it can be generated by TBP with Sulfate ion

$$Th_{aq}^{4+} + 4SO_4^{-}{}_{aq} + nTBP_{org}^{-} \leftrightarrow (Th(SO_4)_4), nTBP)_{org} \qquad (2)$$

From equationn 1 and 2 it can be simplified in the following equation:

$$Th^{4+} + n\ TBP_{org} \rightarrow Th(nTBP)_4$$

Solid phase extraction is one of the important pre-concentration-separation procedures for trace heavy metal ions, due to its simplicity and limited usage of organic solvents. Furthermore, it has many advantages, such as high enrichment factors, absence of emulsion, low cost due to low consumption of reagents, flexibility and being environmentally friendly [8].

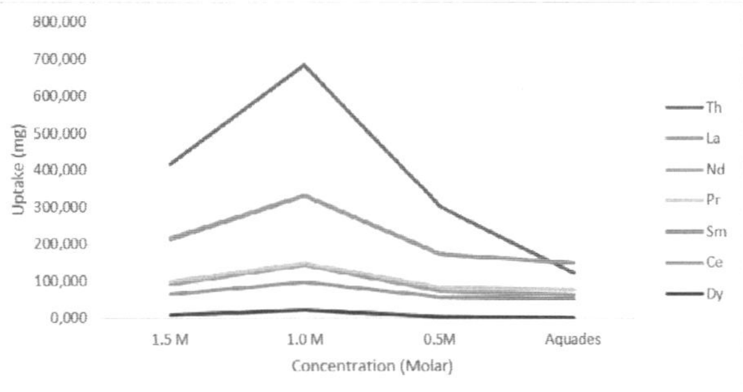

Figure 4. Influence of HNO_3 Concentration

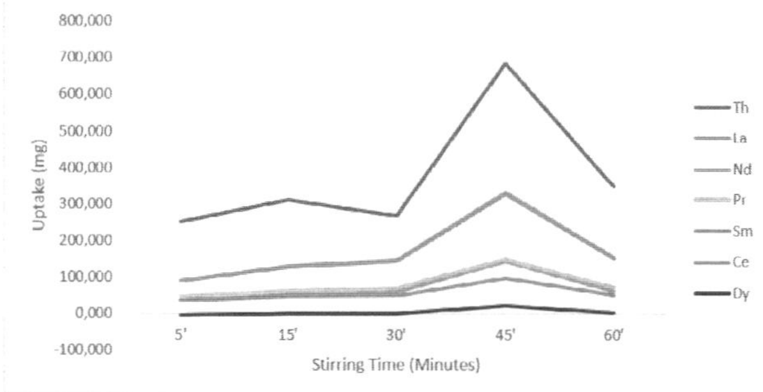

Figure 5. Influence of Stirring Time

3.2 Analysis Effect of solution concentration *(PH Condition)*

By comparison of elution test results between the Rare Earth Elements (REE) element and the Thorium element, as shown in figure 4. there is an increase in elution concentration at Th4+ with a peak of 1.0N HNO3 solvent concentration of 684mg. While in the next concentration after the peak, the elution concentration decreases. So in this process equilibrium point at a concentration of 1.0M is obtained and this process is sufficiently conducted until the concentration of 0.5M - 1.0M HNO3. The elution process may also be applicable to the REE that may be adsorbed by the impregnated resin.

Having the same tendency as Th4 + elution, there is an increase with a peak at a concentration of 1.0M HNO3 with the highest concentration of REE on Neodymium and Cerium elements, but not as many elution results in Thorium, Neodymium produces only 331,25mg and Cerium 327,65mg. In conclusion, of the total mass impregnated by the resin, 1.0M HNO3 solvent has succeeded in eluting each element with the ratio between Thorium and Neodymium of 2,06mg : 1mg. Each 1mg Neodymium will produce 2,06mg element Thorium at the optimum concentration condition.

At higher acidity, the process can be attributed to the competitive adsorption between proton from HNO3 and Th4+. Also at higher concentrations, the reaction of thorium nitrate complex is not perfect. So, it could be resulting the decreased of Th4+ adsorption as HNO3 is increased [7,10, 11]. The increase of adsorption capacity can be attributed to the salting-out effect of HNO3[11]. As it is been studied that elutant HNO3 can act as an absorption or desorption [5]. Based on temporary conclusion that, HNO3 will act contradictively until it is peak concentration as shown in figure 4.

3.3 Analysis *Effect* of Stirring Time

As in figure 5, the fixed variable HNO_3 concentration in the elution process in this experiment is 1.0M. The resulting elution concentration increase at Th^{4+} with a concentration of 684mg is obtained at the peak of stirring time of about 45 minutes. While longer than that, the elution concentration will decrease drastically.

Having the same tendency of elution process on REE has increased with a peak at 45 minutes stirring time. The highest concentration is obtained on the elements of Neodymium as much as 331,25mg and Cerium as much as 327,65mg.

As in figure 5, it is found that the peak is the saturation time of the elution process. Considering HNO3 could act as both adsorption or desorption, It is concluded that the elution process is sufficient for the duration of stirring not more than 30-45 minutes or the elution process will result contrary.

4. Conclusion

An optimum Elution process of Thorium element is obtained as much as 684mg with the stirring condition on 45 minutes at 1.0M HNO3 solvent concentration. The SIR method of the thorium extraction process has selective properties. This process results in 92,432 %recovery, which is equivalent to 684mg of Thorium that is eluted from 740mg of Thorium feed which is contained in the impregnated resin. The lowest elution result is obtained on 5 minutes stirring time at 0M conditions (aquades) with the result of 9.101 %recovery, which is equivalent to 67,350mg. The separation under the optimum conditions of the elution process yielded good results with a very high selectivity of Th4+ element. However, consider the turning point of the solution condition in which the solution will work in reverse. So that the separation process by HNO3 is better done at specific condition.

Acknowledgment

This study was supported by Center for Nuclear Minerals Technology, National Nuclear Energy Agency of Indonesia (BATAN) and Metallurgy & Materials Engineering Department, Universitas Indonesia

References

[1] Burker, T.J., May,1982, *The Characterization of Commercial Thorium Oxide Powders*, Bettis Atomic Power Laboratory, Pennsylvania.

[2] Cuthbert, F.L., 1958, *Thorium Production Technology*, Addison Wesley Publishing Company Inc, USA

[3] Suyanti dan Suprihati, 2010, *Pemurnian Thorium dengan Cara Ekstraksi Memakai Tributil* Fosfat, Prosiding PPI-PDIPTN Pustek Akselerator dan Proses Bahan- BATAN, Yogyakarta

[4] Prassanti, R., 2015, *Pemisahan Uranium pada Monasit dengan Metode Solvent Impregnated* Resin*(SIR).*

[5] Yenes, I., Elif, V.O., Ibrahim,D, et al., 2017, *A new Method for Preconcentration of Th(IV) and Ce(III) by Thermophilic Anoxybacillus Flavithermus Immobilized on Amberlite XAD-16 Resin as a Novel Biosorbent*, Echological Engineering, Elsevier, Turkey

[6] Michaud, S., Miguirditchian, M., Deblonde, G., et al., 2012, *Modelling of Thorium Extraction by TBP*, Procedia Chemistry, Elsevier, France

[7] Patkar, S.N., Burungale, A.S., dan Patil, R.J., 2009, *Separation and Liquid-liquid Extraction of Thorium (IV) as Sulfate Complex with Synegistic Mixture of N-nOcylaniline and Trioctylamne as an Extractant,* Rasayan J. Chem., Deparment of Chemistry, Karnaveer Bhauro Patil College, India. Perry, R.H and Green, D.W., 1999, *Perry's Chemical Engineers Handbook,* McGraw-Hill Book Company, New York.

[8] Rao, T.P., Metilda P, Gladis JM.,2005, Preconcentration techniques for uranium(VI) and thorium(IV) prior to analytical determination-an overview, Talanta 65(1): 1047–1053

[9] Cheira, M.F., 2015, Synthesis of Pyridylazo Resocinol-functionalied Amberlite XAD-16 and Its Characteristics for Uranium Recovery, Journal of Environmental Chemical Engineering, volume 3, Egypt

[10] Sen Qiu, Shun Li, Yamin Dong, Xiang Su, Yanliang Wang, YinglinShen,XiaoqiSun, 2017, A High-performance Impregnated Resin for Recovering Thorium from Radioactive Rare Earth Waste Residue, Journal of Molecular Liquids, China.

[11] P. N. Pathak, R.V. V. K. Manehanda., J. Radioanal. Nucl. Chem,1999, *Separation of Uranium and Thorium Using tris(2-ethylhexyl)phosphate as Extracant.* Journal of Radioanalytical and Nuclear Chemistry,15-18, India

IOP Conf. Series: Materials Science and Engineering **303** (2017) 012006 doi:10.1088/1757-899X/303/1/012006

Finite element simulation and experimental analysis of thermal distribution of optical transceiver

ZHANG Sheng[1, a], NIE Lei[2, b*], JIANG Chuan Kai[3, c]

School of Mechanical Engineering ,Hubei Univ. of Tech. ,Wuhan 430068, China

[a]zhang_sheng9311@163.com

Abstract. In order to optimize the heat dissipation design, the finite element simulation and temperature measurement experiment were used to research the optical transceiver temperature distribution. The results indicated that the shield cage impaired the convective heat transfer efficiency of the photoelectric conversion chip in the optical transceiver. Thus the heat dissipation of the device was weakened. The optimization method was put forward to introduce the external ducts by improving the structure design of the shield cage. The simulation showed the effectiveness of this method which could improve the heat dissipation efficiency of optical transceiver products.

1. Introduction

With the development of optical communication industry which tended toward the miniaturization and integration , the energy density of optical transceiver, which was a typical optical communication device, was getting higher and higher. Heat dissipation now was being a severe challenge for these transceivers and having received great attention from research institutions and industries. Cai proposed two methods to enhance the cooling efficiency of the optical module. One was to open some holes on the PCB board, the other was to fill a thin layer of electric-insulated heat-conducted material between the PCB and package [1-2].And artificial diamonds with high thermal conductivity were embedded in the PCB of the high power devices of optical module, which reduced the difference of temperature between heating part and module shell and achieved better cooling efficiency [3].

These research above provided useful enlightment for the thermal design of optoelectronic devices represented by optical transceivers. However, it was difficult to obtain the internal temperature information directly because the heating device was enclosed inside the packaging. The prototype of optical transceiver products were chose for the research. The simulation was carried out to reveal the internal temperature distribution, found the temperature concentrated area and the route of heat dissipation. After that, the infrared measurement experiment was presented to verify the accuracy of the simulation results and mastered the main form of heat transfer. Finally, based on the conclusion, the optimization measures were put forward.

2. Finite element analysis principle

There were three kinds of heat transfer: radiation, convection and conduction. The heat conduction equation described the internal temperature distribution of the structure. Differential equations could be simplified as follows:

$$pc\frac{\partial T}{\partial t}=\frac{\partial}{\partial x}\left(\lambda_x\frac{\partial T}{\partial x}\right)+\frac{\partial}{\partial y}\left(\lambda_y\frac{\partial T}{\partial y}\right)+\frac{\partial}{\partial z}\left(\lambda_z\frac{\partial T}{\partial z}\right)+\rho q_i\,. \tag{1}$$

Among, p was the material density, c was material specific heat capacity, λx, λy, λz was the heat transfer coefficient x, y, z indirection, qi was the internal chip density, t is time.

The left side of the differential equation represented the amount of heat required per unit time. The first, second and third items on the right side described the heat transfer to the material along the x, y, z three directions per unit time. The last item on the right side was the amount of heat generated of heat source per unit time.

In this paper, the temperature field was isotropic, and the differential equation could be simplified as follows:

$$\frac{\partial^2 T}{\partial x^2}+\frac{\partial^2 T}{\partial y^2}+\frac{\partial^2 T}{\partial z^2}+\frac{\rho q_i}{\lambda}=0\,. \tag{2}$$

The purpose of the study was to find the boundary value of the differential equation. The boundary value studied was a mixed problem of the value of the boundary temperature and the heat transfer between the boundary and the surrounding medium, its mathematical expression was:

$$\lambda\left(\frac{\partial T}{\partial n}\right)_\tau+\alpha(T-T_m)-q_0=0\,. \tag{3}$$

Among, α was heat transfer coefficient; Tm was surrounding medium temperature.

3. Finite simulation analysis

3.1. Design of simulation analysis.

The structure of optical transceiver generally consisted of six parts: the upper and lower shell, the internal structure, the shield cage, the photoelectric conversion chip and the PCB as shown on Fig. 1. The photoelectric conversion chip was packaged on the PCB. The shield cage was used to provide electromagnetic isolation, which was tightly bonded with the PCB by thermosetting polymer. In order to investigate the influence of shield cage on heat dissipation and temperature distribution, two different models were created: with shield cage and without shield cage.

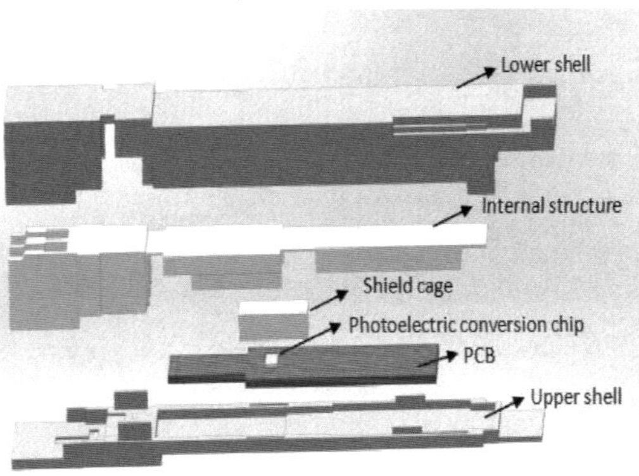

Figure. 1 3D model of the optical transceiver

The power of photoelectric conversion chip was 1W, and the ambient temperature was set to 25 °C. The different material of characteristic parameters were listed in Table 1, such as Young's modulus, thermal conductivity and electrical conductivity.

Table 1 Simulation parameters

Property	Thermal conductivity [W/m.K]	Electrical conductivity [Ω.m]	Young's modulus [N/m^2]	Poisson's ratio
Upper and lower shell, internal structure	237	2.33E-08	70E+09	0.30
Shield cage	214	2.83E-08	70E+09	0.32
Photoelectric conversion chip	15	1.18E-06		
PCB	6.5	1.2E+12		

3.2. Simulation results analysis.

Fig. 2 showed the temperature distribution of the device without shield cage. The chip temperature was the maximum, about 50.8°C. The upper shell surface temperature was about 33.5°C, because of natural convection. Fig. 3 was the cross-sectional temperature distribution of the device. The PCB was in separated state with the load-carrying component, because the intercept plane did not contain support points. The temperature was concentrated on the front of the PCB board, which was gradually decreased from the chip to the middle of PCB and approximately tended to be the same at the end of PCB.

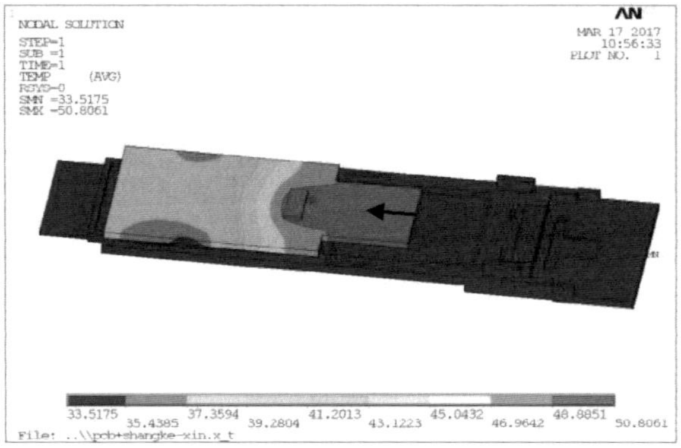

Figure. 2 Temperature distribution without shield cage

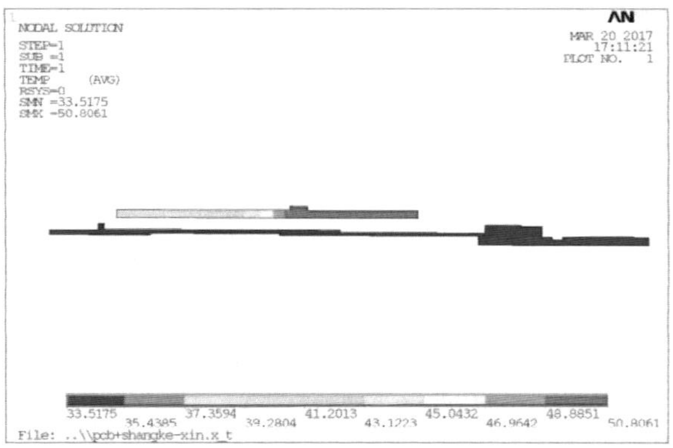

Figure. 3 Device section temperature distribution

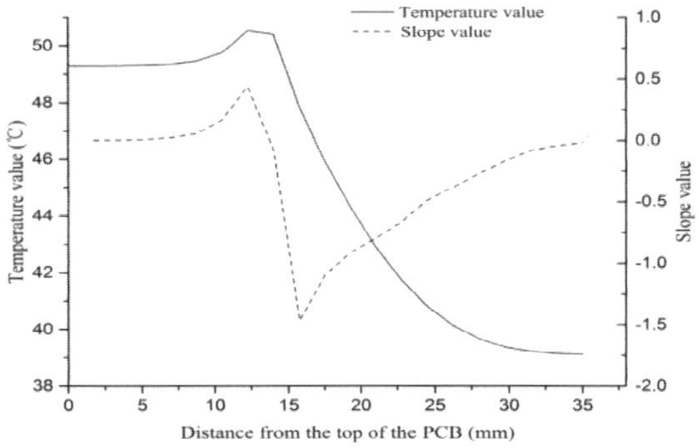

Figure. 4 PCB board temperature distribution

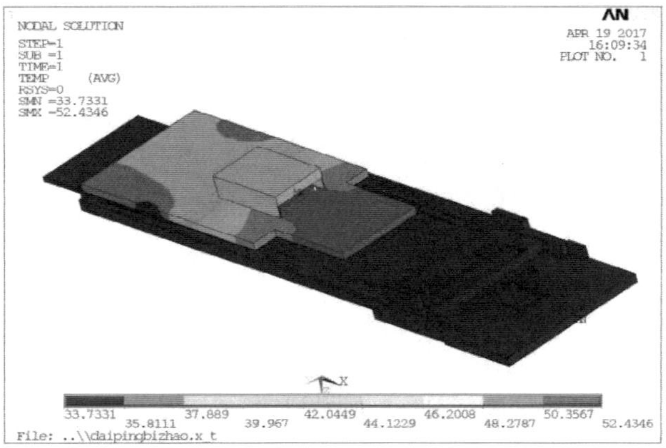

Figure. 5 Temperature distribution with shield cage and slope value

When the PCB board temperature distribution and slope value (the direction of the arrow in Fig. 2) were compared in Fig. 4, it could be found that the slope of the temperature curve of PCB board was almost straight and the position near the chip was a high temperature area and slope value gradually reached its peak value. With the distance increased, the temperature dropped sharply, and the rate of temperature change gradually became slow after the valley value was reached.

The temperature distribution of the device with shield cage was generally similar to Fig. 2 (as shown in Fig. 5). But the maximum temperature of the chip was about 52.4°C, which was 1.6°C higher than that without shield cage. The shield cage temperature was 37°C, and obviously lower than the temperature of PCB. It indicated that the temperature of the PCB board was not effectively transferred to the shield cage. The shield cage hindered the heat dissipation of the chip and increased the temperature.

3.3. Temperature measurement.

A test system was built to verify the validity of the simulation analysis, as shown in Fig. 6. The optical transceiver was placed on the platform, and the infrared imager was mounted on a tripod to record the surface temperature distribution of the device.

IOP Conf. Series: Materials Science and Engineering **303** (2017) 012006 doi:10.1088/1757-899X/303/1/012006

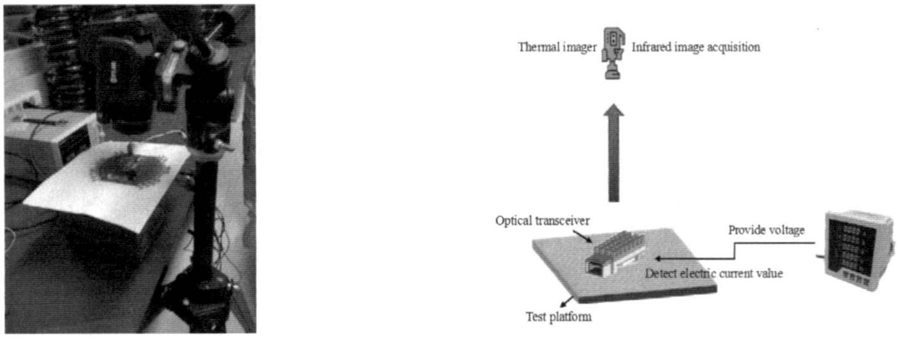

(a) (b)

Figure. 6 Temperature measuring system. (a)Test physical picture, (b) Test sketch picture

From Fig. 7, the temperature distribution without shield cage could be obtained clearly. The brightest area meant the highest temperature, i.e. the photoelectric conversion chip. The brightness at the front of chip indicated the temperature concentrated on this part, and gradually reduced from the front to the end of device. The results were similar to those of Fig. 2.

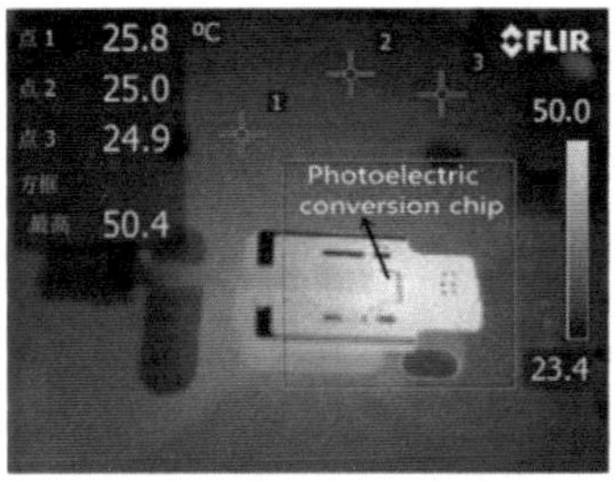

Figure. 7 Infrared temperature distribution image without shield cage

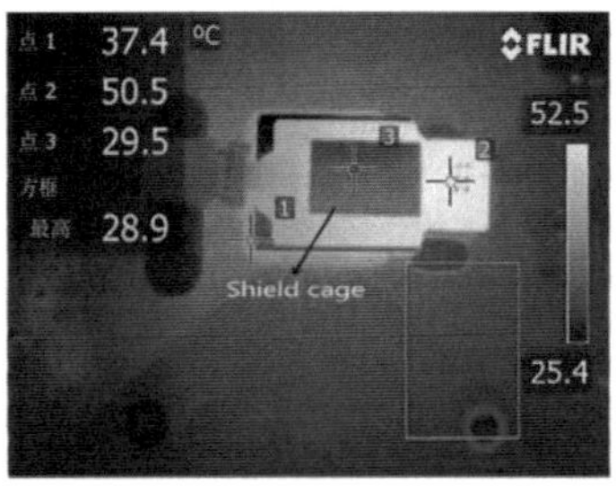

Figure. 8 Infrared temperature distribution image with shield cage

Fig. 8 showed the temperature distribution with shield cage. The highest temperature on the temperature bar approximately represented the chip temperature which couldn't be directly measured due to the presence of the shield cage, it was 52.5℃. The temperature of the shield cage in point 3 was 29.5°C and 20°C lower than that of the chip. The results were similar to those of Fig. 5.

4. Improvement measure

The external air duct was brought to enhance the convection of the chip by alternately punching on the side of the shield cage (as shown in Fig. 9).

IOP Conf. Series: Materials Science and Engineering **303** (2017) 012006 doi:10.1088/1757-899X/303/1/012006

Figure. 9 The improvement shield cage model

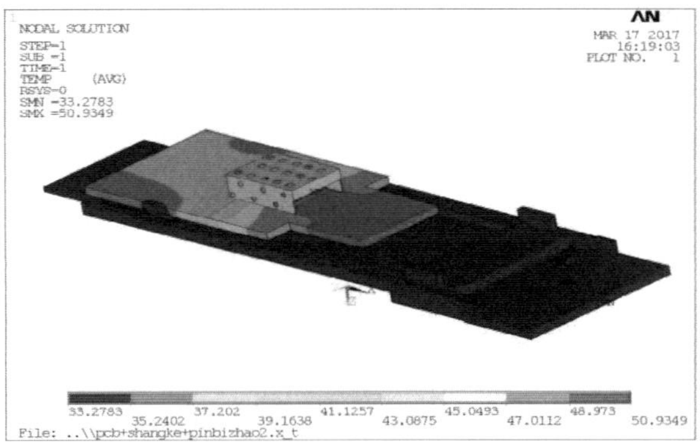

Figure. 10 Simulation of thermal distribution of improvement optical transceiver

Fig. 10 showed the simulation results of the improved optical transceiver. Compared with Fig. 5, the temperature of the chip was decreased nearly 1.4°C, which indicated that the measure was effective after the introduction of the external air duct.

5. Result analysis

The simulation results were credible because the simulative temperature distribution was accordant with the measurement result. Thus conclusions could be drawn as follow.

The temperature distribution was revealed. The temperature of the photoelectric conversion chip was the highest all over the device. And the temperature gradient reached the peak at the middle of the PCB. The temperature of the edge area of PCB near photoelectric conversion chip was almost same with that of the chip and the far-end of PCB was relatively cool and had the same temperature with the outside shell due to the sufficient convection.

The shield cage hindered the heat dissipation of the chip. As a result, the temperature of the chip with shield cage was 2.8°C higher than that without shield cage. It was obvious that the chip were enclosed by shield cage in the shell where the air convection efficiency was very low.

The heat dissipation of heat conduction was inefficient. The contact areas between chip and other parts were so small that the heat generated in the packaging of device couldn't effectively be conducted to the outside.

6. Conclusion

The finite element simulation and temperature measurement had been carried out to reveal the temperature distribution of the optical transceiver. It was found that the existence of the shield cage results in low convective heat transfer, and the temperature was easy to accumulate in the chip. The modified shield cage with additional through holes was proposed and the simulative result proved that it could reduce the temperature to a certain extent.

IOP Conf. Series: Materials Science and Engineering **303** (2017) 012006 doi:10.1088/1757-899X/303/1/012006

Acknowledgement
This work was funded by the Key Project of Educational Commission of Hubei Province of China (D20131407), open Foundation of State Key Lab of Digital Manufacturing Equipment & Technology (DMETKF2014017) and Special Fund in the Public interest of General Administration of Quality Supervision Inspection and Quarantine of P.R.China (201310004).

References

[1] Cai Y Y. The electromagnetic interference shielding and thermal design for high-speed broadband optical module [D]. Southeast University, 2016.
[2] Haghighi E B, Phan T L, Wuttijumnong V, et al. Improved Air Cooling Heat Pipe Based Thermal Solutions for Heat Sinks in Optical Plug Modules[C]// ASME 2015 International Technical Conference and Exhibition on Packaging and Integration of Electronic and Photonic Microsystems Collocated with the ASME 2015, International Conference on Nanochannels, Microchannels, and Minichannels. 2015:V001T09A007.
[3] Li W Q, Wang X F, Ling X, Li X T. The Thermal Structure of The Optical Module [P]. Fu Jian： CN205301638U, 2016-06-08.
[4] Lu J F, Sun F. Heat Radiation Design and Simulation of a Module [J]. Electronic Sci. &Tech, 2013, (08):19-21
[5] Huang S L, Liu Z M, Zhai L Q. Optical Module Cooling Structure and Electronic Products[P]. GuangDong:CN104661487A, 2015-05-27.
[6] Baek Y, Han Y T, Lee C W, et al. Optical components for 100G ethernet transceivers[C]// Opto-Electronics and Communications Conference. IEEE, 2012:218-219.
[7] Zhong D, Yang S, Li S, et al. METHOD AND DEVICE FOR OPTICAL POWER CONTROL: EP 2713528 A1 [P]. 2014.
[8] Cho H, Kapur P, Saraswat K C. Power Comparison Between High-Speed Electrical and Optical Interconnects for Interchip Communication[J]. Lightwave Technology Journal of, 2004, 22(9):2021-2033.
[9] Ke K. Design and Realization of 100Gbit/s CFP Optical Transceiver Module [D]. Fiber Home Technologies Group, 2015.
[10] Shahbaz M, Lean H H, Shabbir M S. High-speed PCB signal integrity analysis based on digital circuit system[J]. Journal of Guilin University of Electronic Technology, 2012, 16(5):2947-2953.

IOP Conf. Series: Materials Science and Engineering **303** (2017) 012007 doi:10.1088/1757-899X/303/1/012007

Study on Resistive Switching Property of Ti Doped Novel NiO Thin Films

Y Li[1]*, G Y Zhao[2], Z B Kou[3], J C Liu[2] and R Zhu[2]

[1]Advanced material analysis center, Xi' an University of Technology, BOX 759#, No.5Jinhua South road, Xi' an, Shaanxi, China 710048
[2]Material science and engineering school, Xi' an University of Technology, Xi' an, Shaanxi, China 710048
[3]Northwest electric power design institute of China power engineering consulting group, No.22 Tuanjie South road, Xi' an, Shaanxi, China 710075

liying@xaut.edu.cn

Abstract. Ti doped nickel oxide thin films have been fabricated by sol-gel dip-coating process using nickel acetate and tetrabutyl titanate as source materials. The effect of the amount of Ti dopant on the surface roughness, optical, chemical state and electrical properties of NiO: Ti thin films was observed by atomic force microscopy (AFM), Uv-vis spectroscopy, X-ray photoelectron spectroscopy(XPS) and I-V measurement, respectively. Results show that the Ti doping is an effective ways to improve the resistive switching behaviors and it is a convenient way to understand the mechanism of the resistive switching behaviors.

1. Introduction

Recently, metal oxides based on resistive switching random access memory (RAM) has attracted great interest as the next generation nonvolatile memory materials.[1-3] Nickel oxides are P type and wide band gap semiconductor as resistive switching materials due to excellent durability, electrochemical stability. [4-6]

One of the effective ways is doping of different dopant, which can improve the resistive switching properties.[7, 8] Additionally doping is a very convenient ways for semiconductors processing combined with sol-gel process. Point defects in oxide films are known to play a critical role in resistive switching mechanism. These defects may be formed in metal oxides by doping. Doping materials is with bonding valence different from that of the constituting elements of the metal oxide with different oxide formation energy. Recent studies show that the electric conductivity of NiO thin film increases by lithium (Li) and copper (Cu) dopants. In our research, we produced a sol-gel process of tetra valence (Ti) atom dopant. [7-8]

2. Experimental

Please The substrate is self-made transparent and conductive SnO2: Sb (ATO) thin film on Si substrate.[9] The ATO substrates were clean by proponal and acetone step by step. Nickelous acetate (Ni(CH3COO)2 • 4H2O) and ethylene glycol monomethyl ether (CH3OCH2CH2OH) were used as starting materials and Tetrabutyl Titanate was used as a dopant. With ethylene glycol monomethyl

ether as solvent, components were mixed in terms of the molar ratio Ni(CH3COO)2 • 4H2O / Ti =1/0.02, 1/0.04, 1/0.06,1/0.08 were stirred for 8 hours. Then a dip-coating method was used to make a gel film on the ATO substrate. After anneal the gel films for 20 minutes at 500C, NiO:Ti thin films were obtained. Then the NiO:Ti thin film was covered by a mask and Pt (99.99%) was evaporated by high vacuum thermal evaporation at 2.0×10-4 Pa for 5 minute at a speed of 10 nm / min. Finally, the unit Pt/ NiO:Ti /ATO was fabricated.
The Ti-doping NiO films were prepared by dip-coating on quartz and ATO substrates respectively. The optical properties of Ti-doping NiO films were examined by UV-vis spectrum and current-voltage characteristics were examined by Keithley 2400c source meter unit. XPS (X-ray photoelectron spectroscopy) and AFM (Atomic force microscopy) were also used for chemical and physical properties' research.

3. Results and Discussion

3.1. Morphological characterization

AFM was used to examine the morphology of the surface of NiO:Ti thin films. To evaluate the surface roughness as well as the grain size in films, a scanned area of a 1um* 1um was scanned in a tapping mode. The two dimensional AFM images were shown in Fig.1. The result shows that all the samples are uniformed and the room meant square (RMS) value is 1.65nm, 1.86nm, 2.23nm and 3.14nm, respectively. The surface roughness increases as a result of Ti dopant concentration increase. This is because the diameter of Ti atom (0.20nm) is bigger than that of Ni atom (0.16nm). When the amount of Ti dopant is increased, the NiO grains are enlarged. This leads to an increase in the grain size and an increase in surface roughness.

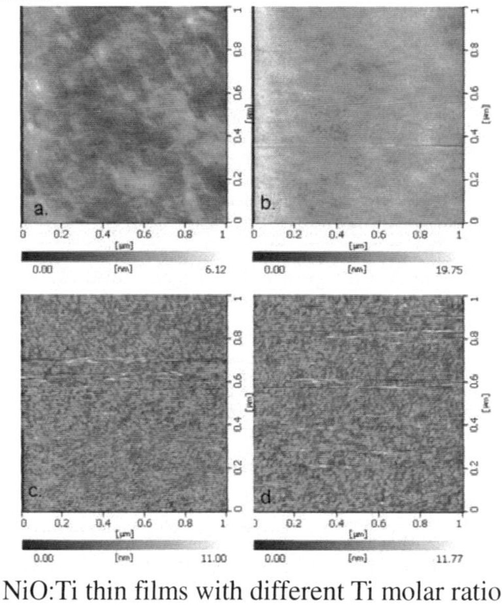

Figure 1. AFM images of NiO:Ti thin films with different Ti molar ratio concentration. (a) 0.02. (b) 0.04. (c) 0.06. (d) 0.08.

IOP Conf. Series: Materials Science and Engineering **303** (2017) 012007 doi:10.1088/1757-899X/303/1/012007

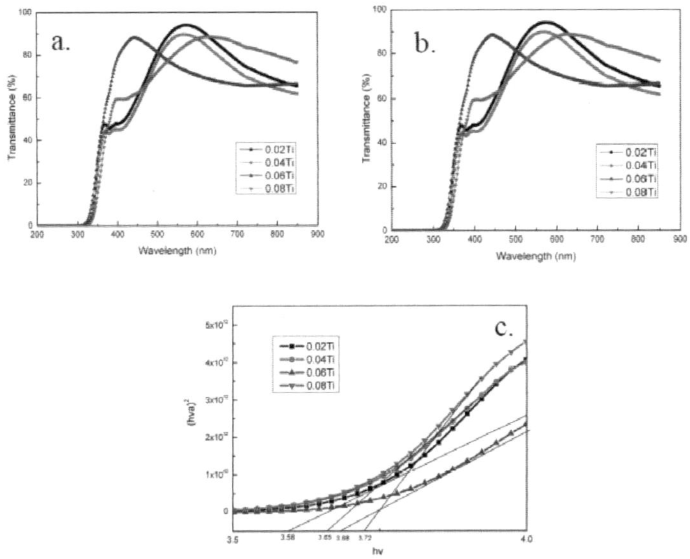

Figure 2. Optical spectra of NiO:Ti films with different Ti molar ratio concentration. (a) Transmittance spectroscopy. (b) Reflectance spectroscopy. (c) Calculated absorption coefficient (α).

3.2. Optical analysis of Ti doped NiO films

Transmittance spectra, reflectance spectra of Ti doped NiO thin films are shown in Fig.2 (a) and (b). The transmittance spectra show that the front edges of the curves represent the intrinsic absorption of nickel oxide. It shows that the films exhibit high transparencies in the visible spectral regions. The highest average transparency is 94.5% when the Ti molar ratio is 0.02. The transmittance of the NiO thin films in the visible region is decreased with the increases of Ti concentration. The optical band gap of Ti doped NiO thin films was calculated using the data of the transmittance and the reflectance versus h ν plot. The absorption coefficient (α) is calculated and showed in Fig. 2(c) by the following Eq.(1) and Eq.(2) :[10,11]

$$\alpha=(1/d)\times \ln[(1-R)/T] \qquad (1)$$

$$\alpha h\nu=C(h\nu-Eg)^{1/2} \qquad (2)$$

where T is the transmittance, R is the reflectance and d is the film thickness. The band gap of Ti doped NiO thin films is determined by plotting (α h ν)2 vs h ν . The calculated optical band gap energy (Eg) decrease from 3.72ev to 3.64ev as increase of Ti dopant concentration which is shown in Table 1.

Table 1. Table of different factors on the increase of Ti dopant concentration.

Ti concentration (%)	Optical band gap (ev)	RMS roughness (nm)	Valence band gap edge (ev)	O1s (1) concentration (%)
2.0	3.58	0.81	3.72	40.2
4.0	3.65	2.33	3.69	42.3
6.0	3.68	2.58	3.66	51.5
8.0	3.72	3.27	3.64	58.7

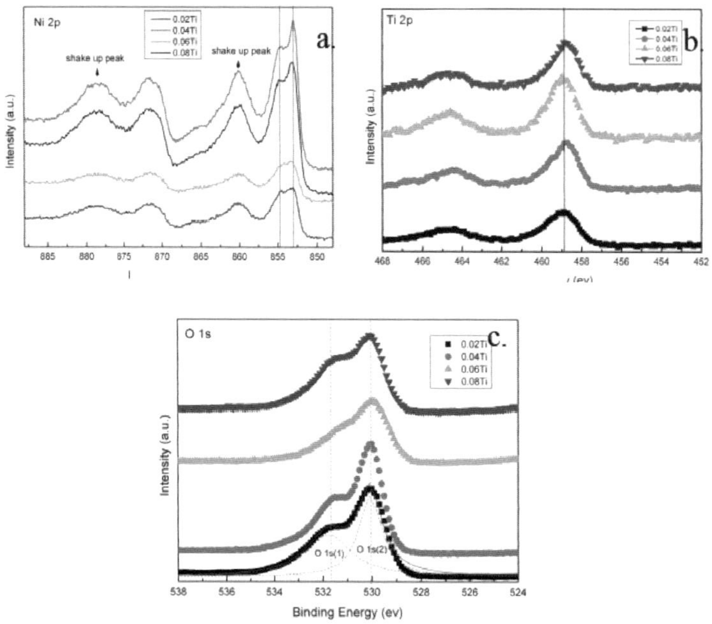

Figure 3. XPS analysis of NiO:Ti thin films with different Ti molar ratio concentration. (a) Ni 3d core level spectrum. (b) Ti 2p core level spectrum. (c) O 1s core level spectrum.

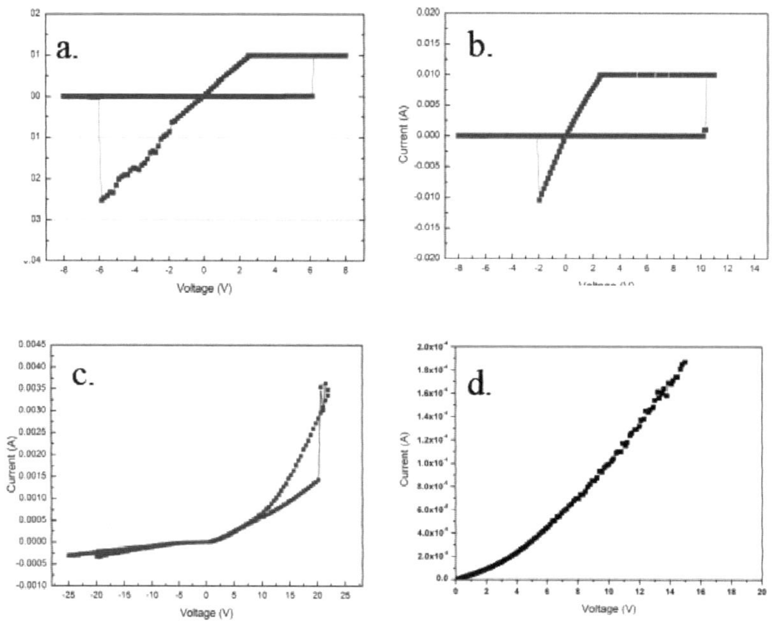

Figure 4. I-V curves of NiO:Ti thin films with different Ti molar ratio concentration. (a) 0.02. (b) 0.04. (c) 0.06. (d) 0.08.

3.3. Characterization of chemical bonding states

The chemical states of the elements in NiO:Ti thin films were measured by XPS. XPS spectra were obtained with monochromatic Al Kα (1486.71eV) line at a power of 150 W (10mA,15kV). The charge neutralizer was applied to compensate for surface charge effects, and binding energies were calibrated using the C1s hydrocarbon peak at 284.8eV. Fig. 3 shows the spectra of Ni3d, Ti2p and O1s core level positions for Ti doped NiO thin films. For Ni3d spectra, two peaks which are labeled at 855.04ev and 856.65ev are represented NiO and Ni2O3, respectively. The basic mechanism leading to the appearance of shake-up satellites are sampling the unoccupied discrete orbital which is also shown in Ni3d spectra. In Fig. 3(b), it is shown Ti2p spectra, which is attributed to bond Ti-O. In Fig. 3(c), two peaks located around 533.61ev (O1s(1)) and 532.04ev (O1s(2)) are attributed to bond Ti-O and Ni-O, respectively.[12] From the results, Ti-O bond component increases obviously with increasing the amount of Ti dopant.

For further research, we found out that the chemical nature differences between Ti and Ni elements played an important role since the Gibbs standard free energy of oxide formation for Ti to form TiO2 is much lower than that for Ni to form NiO. Therefore, it is easier for Ti to bond with oxygen in NiO:Ti thin films and the oxygen vacancy appears easily around the Ti atoms. The oxygen vacancies are associated with the statistically homogeneously distributed Ti ions and the conductive filament consisting of oxygen vacancies accur. [13]

3.4. Electrical measurement

Fig.4 shows I-V curves of Pt/ NiO:Ti /ATO unit cells with different Ti doping concentrations from 20 switching cycles for each unit. It shows that all the units are bipolar except Fig.4 (c). For the switching curves at the beginning of I-V test, a forming process is required. The current compliance (CC) is applied to 10 mA. In Fig. 4 (a) when the voltage goes to 6.0 V, the current increases rapidly and this show the OFF state switches to the ON state. During the second I-V sweep at the negative side, when the voltage goes to -6.0 V, the current decreases rapidly which shows the ON state switches to the OFF state. In Fig. 4 (b) when the voltage goes to 10.0 V, the current increases rapidly and this show the OFF state switches to the ON state. During the second I-V sweep at the negative side, when the voltage goes to -2.0 V, the current decreases rapidly which shows the ON state switches to the OFF state. In Fig. 4 (c) when the voltage goes to 20.0 V, the current increases rapidly and this show the OFF state switches to the ON state. During the second I-V sweep at the negative side, the ON state are not hold any more. These are typical switching behavior of the classical filament type conduction. It shows that the Ti doping can effectively increase the carrier concentration and reduce the resistivity of NiO thin films. It has been proven by Fig. (a) and (b). The low resistive state (LRS) resistance in Fig. (b) is less than that in Fig. (a). However, with increasing the Ti concentration, the LRS resistance is increased when the Ti molar ratio concentration is increased to 0.06 and that means the Ti doping is saturated. In Fig. 4 (d), it is a typical semiconductor I-V curve.

4. Summary

Sol-gel process is a convenient ways for doping. The surface roughness (RMS) is influenced by doping when the Ti dopant concentration is increased while the RMS value is also increased. The Eg is decreased when the amount of Ti dopant is increased. The chemical state of Ti dopant is studied by XPS. It shows that Ti exits in Ti-O bond. O1s (1) in Ti-O bond is increased when the amount of Ti dopant is increased. On the other hand, O1s (2) in Ni-O bond is decreased when the amount of Ti dopant is increased. When the concentrations of Ti come to 0.06, Ti doping is saturated. I-V curves shapes are bipolar except when the concentration of Ti is 0.08.

5. Acknowledgements

This work is supported by the National Natural Science Foundation of China (No. 51372198). We thank the Foundation of young teachers of Xian University of Technology.

6. References

[1] Zhou P.,Shen H.,Li J.,Chen LY.,Gao C., Lin Y. ,T.A.Tang. Resistance switching study of stoichiometric ZrO_2 films for non-volatile memory application [J].Thin Solid Films 2010,518,5652–5655.

[2] Cao X., Li X.,Yu W.,Zhang Y., Yang R., Liu X., Kong J., Shen W. Structural characteristics and resistive switching properties of thermally prepared TiO_2 thin films [J].Journal of Alloys and Compounds, 2009, 486, 458–461.

[3] Cao X., Li X.,Yu W., Zhang Y. Preparation of PLD TiO_2 thin films and study on the resistance switching properties of TiO_2 thin films [J]. Journal of Inorganic Materials, 2009, 24(1):49-52.

[4] Min K.-C., Kim M., You Y.-H., Lee S.S., Lee Y.K., Chung T.-M., Kim C.G.,Hwang J.-H., An K.-S., Lee N.-S., Kim Y. .NiO thin films by MOCVD of $Ni(dmamb)_2$ and their resistance switching phenomena [J]. Surface & Coatings Technology, 2007, 201, 9252–9255.

[5] Courtade L., Turquat Ch., Muller Ch., Lisoni J.G., Goux L., Wouters D.J., Oguenheim D.G , Roussel P., Ortega L,.Oxidation kinetics of Ni metallic films: Formation of NiO-based resistive switching structures [J]. Thin Solid Films, 2008, 516, 4083–4092.

[6] Kügeler C., Weng R., Schroeder H., Symanczyk R., Majewski P., Ufert K.-D., Waser R., Kund M. Study on the dynamic resistance switching properties of NiO thin films [J]. Thin Solid Films, 2010, 518, 2258–2260.

[7] Sta I., Jlassi M., Hajji M., Ezzaouia H.. Structural, optical and electrical properties of undoped and Li-doped NiO thin films prepared by sol-gel spin coating method. Thin solid films. 555(2014)131-137.

[8] Zhao L., Su G., Liu W. et al. Optical and electrochemical properties of Cu-doped NiO thin films by electrochemical deposition. Applied surface science. 257(2011)3974-3979.

[9] Zhi X., Zhao G., Zhu T., LibY.. Surf. Interface Anal. 40(2008), 67-70.

[10] Lee D-Y. , Kim J-T., Park J-H., et al. Effect of Er doping optical band gap energy of TiO_2 thin films prepared by spin coating. Current applied physics. 13(2013) 1301-1305.

[11] Chen X., Song L., You L., Zhao L.. Incorporation effect of Y_2O_3 on the structure and optical properties of HfO_2 thin films. Applied surface science. 271(2013)248-252

[12] Moulder J. F. et al, Handbook of X-ray photoelectron spectroscopy, Perkin. 1992.

[13] Paul J.D., Saravanan M., Muthuraaman B..et al. Spray deposition and characterization of nanostructured Li doped NiO thin films for application in dye-sensitized solar cells, Nanotechnology. 19(2008) 485707-485717.

IOP Conf. Series: Materials Science and Engineering **303** (2017) 012008 doi:10.1088/1757-899X/303/1/012008

Estimation of Ultimate Tensile Strength of dentin Using Finite Element Analysis from Endodontically Treated Tooth

S Sinthaworn[1,3], U Puengpaiboon[2], N Warasetrattana[3] and S Wanapaisarn[4]

[1]Research Unit in Sustainable Innovation in Civil and Environmental Engineering, Srinakharinwirot University, Thailand
[2]Department of General Dentistry, Faculty of Dentistry, Srinakharinwirot University, Thailand
[3]Civil and Environmental Engineering Department, Faculty of Engineering, Srinakharinwirot University, Thailand
[4]Department of Conservative Dentistry, Faculty of Dentistry, Western University, Thailand

suppachai@g.swu.ac.th

Abstract. Endodontically treated teeth were simulated by finite element analysis in order to estimate ultimate tensile strength of dentin. Structures of the endodontically treated tooth cases are flared root canal, restored with different number of fiber posts {i.e. resin composite core without fiber post (group 1), fiber post No.3 with resin composite core (group 2) and fiber post No.3 accessory 2 fiber posts No.0 with resin composite core (group 3)}. Elastic modulus and Poisson's ratio of materials were selected from literatures. The models were loaded by the average fracture resistances load of each groups (group 1: 361.80 N, group 2: 559.46 N, group 3: 468.48 N) at 135 degree angulation in respect to the longitudinal axis of the teeth. The stress analysis and experimental confirm that fracture zone is at dentin area. To estimate ultimate tensile strength of dentin, trial and error of ultimate tensile strength were tested to obtain factor of safety (FOS) equal to 1.00. The result reveals that ultimate tensile strength of dentin of group 1, 2, 3 are 38.89, 30.96, 37.19 MPa, respectively.

1. Introduction

Finite element analysis (FEA) is a computational technique widely used to estimate the behavior of structures under loading. It is particularly valuable to investigate those conditions which cannot be studied directly by experiment, and therefore, provides valuable information that cannot be obtained in more direct ways. This technique uses a series of small elements which are connected at nodal points, to computer-generate a morphological structure. The elements can be assigned properties that are indicative of the tissue properties that small areas of the biological structure have, either relying on measurements that have been made of the properties of the material, or in those cases in which the properties have not been measured, estimating them based on the available data [1].

The physical properties of tooth which is highly mineralized tissues based on their composition and micro morphology [3]. Enamel, dentin and dentin–enamel junction are parts of tooth with different strength. However, most of the tooth structure is consisted of dentin which is a hydrated biological composite composed of 70% inorganic material, 18% organic matrix and 12% water (wt%), with properties and structural components that vary with location [4]. Recently, the experiment on

endodontically treated teeth reveals that the fracture normally occurs in dentin area as a weak point [2]. Although, the microtensile technique is a good tool to determine the ultimate tensile strength (UTS) of both materials and tooth substrates [3], alternatively, finite element technique could be used indirectly to estimate the UTS of tooth from result of non-conformed standard of fracture resistance test.
The aim of this research is to estimate ultimate tensile strength of dentin from existing experimental data in dentistry [2] by finite element analysis.

2. Methodology

2.1. Existing Experimental Data
In 2015, Puengpaiboon et al. conducted the research which is determination of fracture resistance of endodontically treated teeth with flared root canal, restored with different number of fiber posts [2]. The samplings includes 3 groups; group 1 was restored with a resin composite core material (Multicore Flow), group 2 was restored with a single fiber post and resin composite core (FRC Postec Plus No.3, Multicore Flow), and group 3 was restored with 3 fiber posts and resin composite core (1 FRC Postec Plus No.3 and 2 fiber posts No.0, Multicore Flow) as shown in Figure1.
All Samples were tested at 135 degree angulation in respect to the longitudinal axis of the teeth using a Universal testing machine as shown in Figure2 (a). The average fracture resistances were as follows: group 1: 361.80 ± 93.16 N, group 2: 559.46 ±155.12 N, group 3: 468.48±155.17 N [2]. The facture of specimens are illustrated in Figure2 (b).

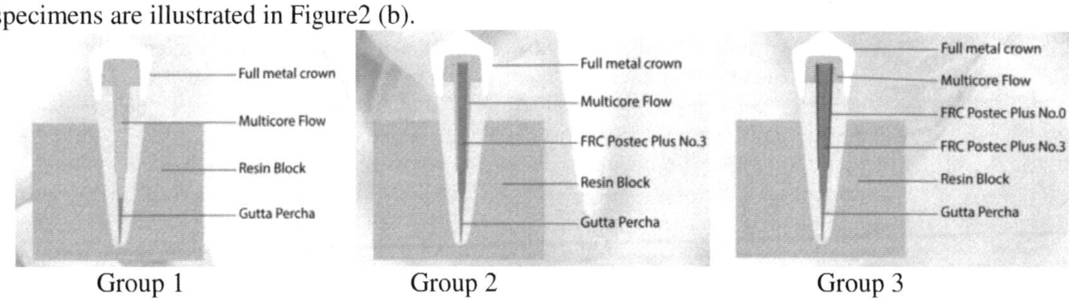

Figure 1. The existing tested samplings [2].

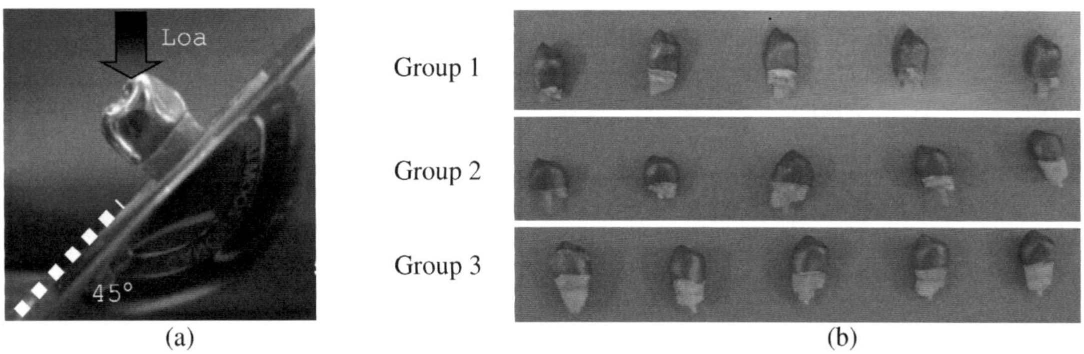

Figure 2. (a) Test direction, (b) facture of specimens of group 1, 2, 3 [2].

2.2. Finite Element Model and Analysis
The simulation are based on linear elastic, isotropic and homogenous material. Dimension of mandibular single-root premolars was obtain from CT scan (approximately length is 22 mm. and 9 mm. in diameter). Simplified geometry of endodontically treated teeth with flared root canal, tooth

IOP Conf. Series: Materials Science and Engineering **303** (2017) 012008 doi:10.1088/1757-899X/303/1/012008

model and full models show in Figure3. Elastic modulus and Poisson's ratio of materials used in the model are shown in Table 1.
Estimation of ultimate tensile strength of dentin was done by trial & error from analysis factor of safety (FOS) until maximum of FOS is one. The critical control stress used in this trial & error is principal tensile stress and the limit strength of material is ultimate tensile strength of dentin as shown the flow chart in Figure4. In addition, from preliminary analysis, ultimate tensile strength of other parts are quite higher than maximum principal tensile stress at that parts. In other words, FOS of other parts is always higher than one. Therefore, fracture part in located only in dentin.

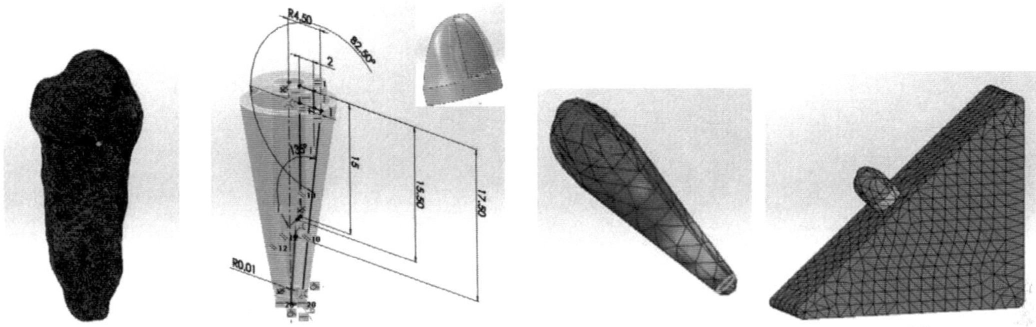

(a) (b) (c) (d)

Figure 3. (a) CT scan of mandibular single-root premolars, (b) simplified geometry of endodontically treated teeth with flared root canal, (c) model of the endodontically treated teeth tooth, (d) full model with resin support.

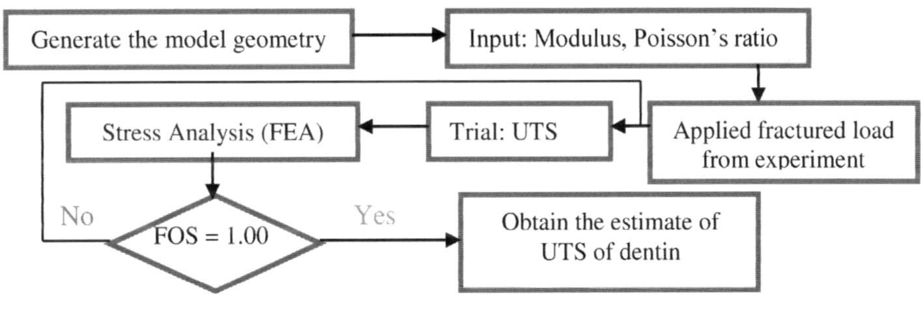

Figure 4. Flowchart to estimate UTS of dentin.

Table 1. Elastic modulus and Poisson's ratio of materials

Material	Young's Modulus (E) MPa	Poisson's ratio	Reference
Dentin	18,600	0.31	[5]
Gutta percha	0.69	0.45	[6]
Resin Composite	16,600	0.24	[7]
Metal Crown	96,600	0.35	[6]
FRC Post	21,000	0.30	[8]
Resin (Block)	20,000	0.30	[9]

IOP Conf. Series: Materials Science and Engineering **303** (2017) 012008 doi:10.1088/1757-899X/303/1/012008

3. Results and Discussion

3.1. Group 1 (restored with a resin composite core material)

Figure 5 (a) displays the model's components which are metal crown, dentin, resin composite, gutta percha and resin support block. Principal stress of endodontically treated teeth group 1 is demonstrated in Figure 5 (b). The maximum compressive stress located in metal crown whereas the maximum tensile stress occurs in dentin. Factor of safety (FOS) are defined as stress divided by strength of material at each point in model. If principal tensile stress caused by external load equals to ultimate tensile strength of material, FOS will be unity. To obtain FOS equals to one, tensile strength of dentin was applied as 38.89 MPa to get FOS in Figure5 (c)

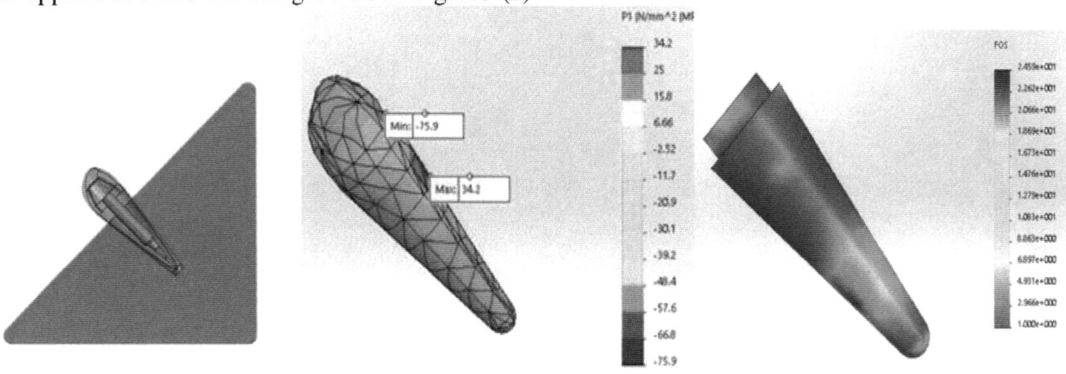

(a) (b) (c)

Figure 5. Group 1, (a) full part of model, (b) principal stress of endodontically treated teeth (c) FOS.

3.2. Group 2 (restored with a single fiber post and resin composite core)

The model's components (metal crown, dentin, resin composite, fiber post, gutta percha and resin support block) are illustrated in Figure6 (a). Principal stress of endodontically treated teeth group 2 is demonstrated in Figure 6 (b). The stress distribution proves the critical principal stress distribute into the root of tooth, deeper than group 1. It might cause of fracture site at middle 1/3 of the root as reported in Puengpaiboon et al. (2015) [2]. Trial of ultimate tensile strength to get maximum FOS to be one is 30.96 MPa results in the analysis of FOS as shown in Figure6 (c).

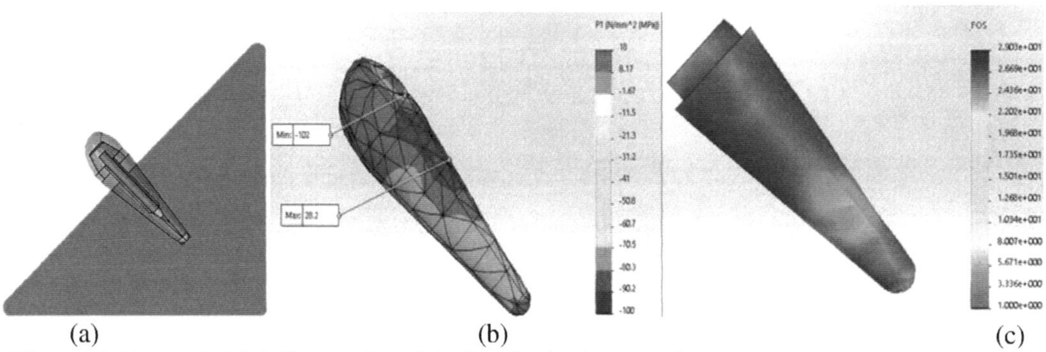

(a) (b) (c)

Figure 6. Group 2, (a) full part of model, (b) principal stress of endodontically treated teeth (c) FOS.

3.3. Group 3 (restored with 3 fiber posts and resin composite core)

Figure 7 displays the model's components, principal stress of endodontically treated teeth (only dentin part is presented), and successfully trial FOS diagram. The depth of critical stress distribution is more or less than that of group 2. It might cause of fracture area similar to group 2 [2]. Estimation of ultimate tensile strength of dentin from this group is 37.19 MPa. In summary, the estimation of ultimate tensile strength of dentin from this research are ranged from 31.0 to 38.9 MPa. On the other hand, Giannini et al. [3] found that ultimate tensile strength of dentin is in the range of 33.9 to 48.7 MPa by microtensile technique of real teeth. In addition, Lertchirakarn et al. found that UTS was lowest (42.0 MPa) when force loading parallel and highest (52.9 Mpa) when angle of loading perpendicular to tubule orientation by diametral test [10]. This work reveals that FEA technique could be used to estimate the ultimate tensile strength of tooth from the existing experimental data.

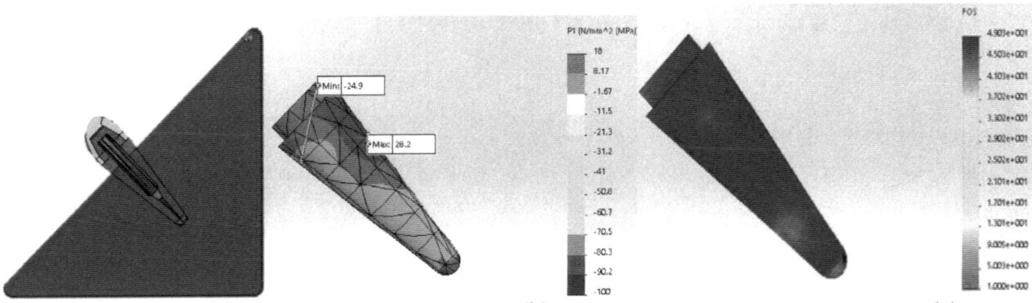

(a) (b) (c)

Figure 7. Group 3, (a) full part of model, (b) principal stress of endodontically treated teeth (c) FOS.

4. Conclusions

The estimation of ultimate tensile strength of dentin from this research are ranged from 31.0 to 38.9 MPa. This work reveals that FEA technique could be used to estimate the ultimate tensile strength of tooth from the existing experimental data.

5. Acknowledgements

The authors would like to thanks Mechanical Engineering Department, Faculty of Engineering, Srinakharinwirot University for SolidWorks Software.

6. References

[1] Burr DB. The use of finite element analysis to estimate the changing strength of bone following treatment for osteoporosis. Osteoporosis International. 2016;27(9):2651–4.

[2] Puengpaiboon U, Didron P P, Vetviriyakul N. comparison of fracture resistance of endodontically treated teeth with flared root canal, restored with different number of fiber posts. Srinakharinwirot University Journal of Science and Technology. 2015:7(13):76-87.

[3] Giannini M, Soares CJ, De Carvalho RM. Ultimate tensile strength of tooth structures. Dent Mater. 2004;20(4):322–9.

[4] Mjör IA. Human coronal dentine: Structure and reactions. Oral Surgery, Oral Med Oral Pathol. 1972;33(5):810–23.

[5] Reinhardt RA, Pao YC, Krejci RF. Periodontal ligament stresses in the initiation of occlusal traumatism. J Periodontal Res. 1984;19(3):238–46.

[6] Asmussen E, Peutzfeldt A, Heitmann T. Stiffness, elastic limit, and strength of newer types of endodontic posts. J Dent. 1999;27(4):275–8.

[7] Rees JS, Hammadeh M, Jagger DC. Abfraction lesion formation in maxillary incisors, canines and premolars: a finite element study. Eur J Oral Sci. 2003;111(2):149–54.

[8] Sirimai S, Riis DN, Morgano SM. An in vitro study of the fracture resistance and the incidence ofvertical root fracture of pulpless teeth restored with six post-and-coresystems. J Prosthet Dent. 1999;81(3):262–9.

[9] Information on http://www.engineeringtoolbox.com/polymer-properties-d_1222.html

[10] Lertchirakarn V, Palamara JEA, Messer HH. Anisotropy of Tensile Strength of Root Dentin. J Dent Res [Internet]. 2001;80(2):453–6. Available from: http://journals.sagepub.com/doi/10.1177/00220345010800021001

IOP Conf. Series: Materials Science and Engineering **303** (2017) 012009 doi:10.1088/1757-899X/303/1/012009

Impact of Aggregates Size and Fibers on basic Mechanical Properties of Asphalt Emulsion—Cement Concrete

Jun Fu[1,*], Zhihong Liu[1] and Jie Liu[1]

[1]School of Transportation, Wuhan University of Technology, Wuhan 430063, China

*fjgrant@whut.edu.cn

Abstract. Asphalt Emulsion—Cement Concrete (AECC) is currently considered as a typical semi-flexibility material. One of the disadvantages of this material is brittle fracture and lacking ductility. This study aims at accelerating the basic mechanical properties of AECC using fibers and different aggregates size. The mix of AECC was introduced and the different content of fibers and aggregates size were studied. The results showed that the smaller aggregates size could improve the young's modulus and compressive strength as well as fiber. The modulus-compressive strength ratio of fiber reinforced AECC is always below 500.

1. Introduction

Cement-Asphalt Emulsion Mixture (CAEM) is made of by Portland cement, emulsified asphalt, aggregates and some admixtures. CAEM is currently considered as easily cool-mixed and cool-spread, so AECC have environmental, economical, and logistical advantages over Hot Mixture Asphalt (HMA)[1].Research on these composite materials, also called Cement-Asphalt Emulsion Composite (CAEC)[2,3], Cold Bituminous Emulsion Mixture (CBEM)[4,5], Cement-Emulsion Treated Mixture (C-ETM)[6] or Cement–Bitumen Treated Material (CBTM)[7]. According to different mix design method, CAEM could be classified with CAEC and AECC. CAEC is always designed with grading aggregates and cold roller compacted in pavement while AECC is casting without accurate grading aggregates.

Some research has also been carried out on emulsified asphalt cement concrete [8,9], in which cement is the primary binder and bitumen an additive. Cold Bituminous Emulsion Asphalt (CBEA) shares characteristics of both cement and asphalt concrete [10,11]. In particular, compared with conventional HMA, CBEA has higher deformation resistance and lower temperature susceptibility [12]. Additionally, the introduction of cement in cold mix asphalt accelerates emulsion breaking [13], because cement hydration consumes water in the emulsion and meanwhile it increases its alkalinity.

However, AECC is still brittle fracture and lacking ductility. This study focused on accelerating the basic mechanical properties of AECC using fibers and different aggregates size.

2. Material and Specimen

2.1 Material

The material used in this study is aggregates, emulsified asphalt, Portland cement, PVA fiber, fly ash, water and additives. The slow-cracking anionic emulsified asphalt is used in this study and the character is given in Table 1. The properties of Portland cement are shown in Table 2. The fly ash content and properties tested is shown in Table 3. The aggregates are limestone from local and the test results is shown in Table 4. The mineral filler is from limestone whose size is smaller than 0.6 mm after sieving. Water reducer agent is early strength Poly carboxylic acid and water meets concrete water standards. Figure 1 and Figure 2 are PVA fiber and fly ash respectively.

Table 1. Emulsified Asphalt Characteristics.

properties	specification limitation	Property index
Sieve residue(1.18mm)/%	<0.1	0.04
particle diameter/μm	average particle diameter≤7 Module size≤5	average particle diameter:6.2 Module size:4.3
adaptability to cement	≥70	100
Residual content/%	≥60	60.8
Penetration (25°C、100g,5s /0.1mm)	40~120	65
Solubility(chlorylene) /%	≥42	46
ductility(25°C、5cm/min)/cm	≥100	126

Table 2. Portland cement properties.

properties	specific surface area(m2/kg)	initial setting time(min)	final setting time(min)	compressive strength(Mpa) 28d	Bending strength(Mpa) 28d
P.O 42.5	310	118	360	45.7	7.5

Table 3. Fly ash content and properties(%).

SiO_2	Al_2O_3	Fe_2O_3	SO_3	CaO	others	Ignition loss	Water demand ratio	Compressive strength ratio
57.67	28.31	3.92	0.43	2.86	1.86	4.95	90	75

Table 4. Aggregates properties.

properties	apparent density (g/cm3)	water absorption (%)	needle-like particles (%)	crushing value (%)	liquid limit (%)	plasticity index	uniformity coefficient
value	2.75	2.5	3.2	12	30	10	13

Figure 1. PVA fiber.

Figure 2. Fly ash.

2.2 Specimen

IOP Conf. Series: Materials Science and Engineering **303** (2017) 012009 doi:10.1088/1757-899X/303/1/012009

After initial determination of the mix ratio of emulsified asphalt concrete, the above mix ratio is used as the basis ratio, and the influence of the particle size and PVA fiber on AECC is analyzed emphatically. The basic mix design and variable parameters are shown in table 5 and table 6. Specimen and mix are shown in Figure 3 and Figure 4. Emulsified asphalt is organic cementitious materials and cement is inorganic cementitious materials, which both form a solid, flexible and rigid integral composite material.

Table 5. Basic mix parameters（kg/m^3）.

Water	Cement	Emulsified asphalt	Fly ash	sands	Coarse aggregate	Fiber	Water reducing agent	Defoamer
115	217	102	38	1089	730	0.9	14	0.012

Table 6. variable parameters.

Fiber（kg/m^3）				Aggregates size （mm）		
0.5	0.7	0.9	1.1	5-10	5-15	5-20

Figure 3. Specimen of AECC.

Figure 4. Mix of AECC.

3. Results and Discussion

As shown in Table 7 and Figure 5, the compressive strength and young's modulus of AECC is negative correlation with the aggregate size，so the small aggregate size is beneficial to improving the strength and young's modulus of AECC. Comprehensive consideration of economic cost and performance, the aggregate of 5-15mm is a relative optimization. The effect of fiber content in AECC is significant from Table 8 and Figure 6. In different age of AECC, with more fiber content, the compressive strength and young's modulus is increased. But in the process of mix, the fiber should be controlled to avoid cluster and preventing the agglomeration. As for modulus-compressive strength ratio, the value for fiber reinforced AECC always be less than 500, which means plastic and suitable ductility to some degree.

Table 7. Test results of concrete with different particle size.

Number	particle size（mm）	Compressive strength（Mpa）			Elastic modulus(Gpa)			Modulus-strength ratio		
		14d	28d	90d	14d	28d	90d	14d	28d	90d
Xq1	5-10	11.6	13.1	18.4	2.25	4.27	7.57	194.0	326.0	411.4
Xq2	5-15	10.5	12.2	17.2	2.13	4.15	7.46	202.9	340.2	433.7

Xq3	5-20	9.5	11.4	16.3	1.97	4.02	7.32	207.4	352.6	449.1

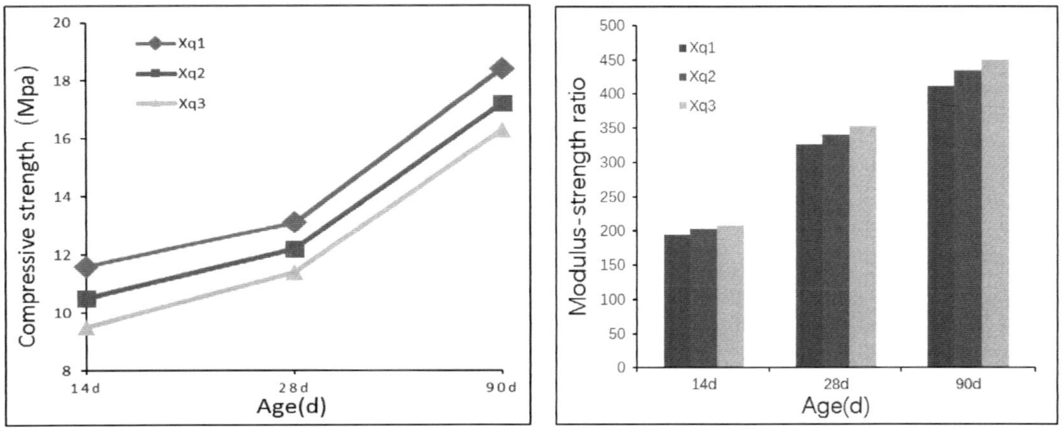

Figure 5. Effect of different particle size on compressive strength and modulus-strength ratio of concrete.

Table 8. Test results of concrete with different fiber content.

Number	Fiber content kg/m^3	Compressive strength (Mpa)			Elastic modulus(Gpa)			Modulus-strength ratio		
		14d	28d	90d	14d	28d	90d	14d	28d	90d
Xq4	0.5	7.1	8.9	13.1	1.75	3.78	7.08	246.5	424.7	540.5
Xq5	0.7	8	10.1	14.2	1.86	3.87	7.17	232.5	383.2	504.9
Xq6	0.9	8.9	11	15.4	1.96	3.99	7.27	220.2	362.7	472.1
Xq7	1.1	9.8	11.8	16.6	2.07	4.11	7.36	211.2	348.3	443.4

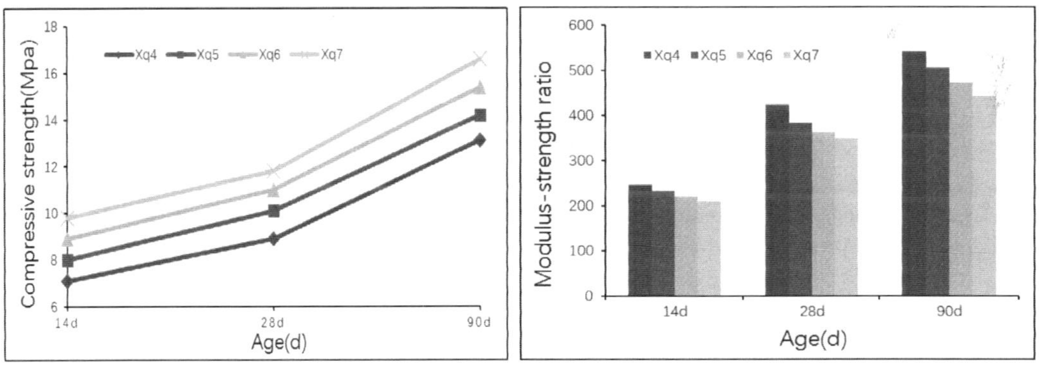

Figure 6. Effect of different fiber content on compressive strength and modulus-strength ratio of concrete.

3. Conclusions

Fiber reinforced AECC is middle and low strength concrete. The compressive strength and elastic modulus will increase when aggregates size is getting smaller. And the PVA fiber can raise the compressive strength and elastic modulus for AECC. Especially, with smaller modulus-compressive strength ratio, the fiber reinforced AECC is better in deformation and ductility than normal Cement Concrete.

4. Acknowledgements

The authors would like to acknowledge support by the National Natural Science Foundation of China (Nos.51072149). Additional Supported by Chinese Department of Transportation Planning Project of Science and Technology (Nos.2015319811320).

5. References

[1] Effect of Cement on Emulsified Asphalt Mixtures. Seref Oruc, Fazil Celik, and M. Vefa Akpinar. JMEPEG (2007) 16:578–58

[2] G. Li, Y. Zhao, S.-S. Pang, and W. Huang, Experimental study of cement-asphalt emulsion composite, 26 Cem. Concr. Res., vol. 28, no. 5, pp. 635–641, 1998. 27

[3] S. Oruc, F. Celik, and M. V. Akpinar, Effect of cement on emulsified asphalt mixtures, J. Mater. Eng. 28 Perform., vol. 16, no. October, pp. 578–583, 2007. 29

[4] H. Al Nageim, S. F. Al-Busaltan, W. Atherton, and G. Sharples, A comparative study for improving the mechanical properties of cold bituminous emulsion mixtures with cement and waste materials, Constr. 1 Build. Mater., vol. 36, pp. 743–748, 2012.

[5] I. N. A. Thanaya, Improving the performance of cold bituminous emulsion mixtures (CBEMs) : 7 incorporating waste, PhD thesis, The University of Leeds, 2003.

[6] R. J. Schmidt, L. E. Santucci, and L. D. Coyne, Performance characteristics of cement modified asphalt 30 emulsion mixes, Proc. Association of Asphalt Paving Technologists, 1973, pp. 300–319.

[7] M. Bocci, A. Grilli, F. Cardone, and A. Graziani, A study on the mechanical behaviour of cement-1 bitumen treated materials, Constr. Build. Mater., vol. 25, no. 2, pp. 773–778, 2011. 2

[8] H. A. Khalid and O. K. Monney, Moisture damage potential of cold asphalt, Int. J. Pavement Eng., vol. 11 10, no. November 2014, pp. 311–318, 2009.

[9] Jun Fu, Xiaoqiang Zhang, and Fazhou Wang.The Meso Structure and Strain Distribution Analysis of Cement Emulsified Asphalt Concrete. International Journal of Pavement Research and Technology,2015,8(3):179-184

[10] M. N. Partl, Sustainable development through asphalt recycling in Switzerland, in International 15 Workshop ISAP Technical Committee APE Asphalt Pavements and Environment, 2010.

[11] A. D. James, D. Needham, and P. S. F. Brown, The benefits of using Ordinary Portland Cement in 17 solvent free dense graded bituminous emulsion mixtures, Proc, International Symposium on Asphalt 18 Technology, 1996.

[12] S. Al-Busaltan, H. Al Nageim, W. Atherton, and G. Sharples, Mechanical Properties of an Upgrading 3 Cold Mix Asphalt Using Waste Materials, J. Mater. Civ. Eng., no. December, p. 463, 2012.

[13] J. Chehovits and L. Galehouse, Energy usage and greenhouse gas emissions of pavement preservation 19 processes for asphalt concrete pavements, Proc. 1st International Conference on Pavement Preservation, 20 2010, pp. 27–42.

IOP Conf. Series: Materials Science and Engineering **303** (2017) 012010 doi:10.1088/1757-899X/303/1/012010

Preparation and Physical Properties of Segmented Thermoelectric $YBa_2Cu_3O_{7-x}$-$Ca_3Co_4O_9$ Ceramics

P Wannasut[1,2], N Keawprak[3], P Jaiban[4],and A Watcharapasorn[1,5]

[1]Department of Physics and Materials Science, Faculty of Science, Chiang Mai University, Chiang Mai 50200, Thailand
[2]Graduate School, Chiang Mai University, Chiang Mai, 50200, Thailand
[3]Thailand Institute of Scientific and Technological Research, Pathum Thani 12120, Thailand
[4]Faculty of Science, Energy and Environment, King Mongkut's University of Technology North Bangkok, Rayong Campus, Rayong 21120, Thailand
[5]Materials Science Research Center, Faculty of Science, Chiang Mai University, Chiang Mai 50200, Thailand

anucha@stanfordalumni.org

Abstract. Segmented thermoelectric ceramics are now well known for their high conversion efficiency and are currently being investigated in both basic and applied energy researches. In this work, the successful preparation of the segmented thermoelectric $YBa_2Cu_3O_{7-x}$-$Ca_3Co_4O_9$ (YBCO-CCO) ceramic by hot pressing method and the study on its physical properties were presented. Under the optimum hot pressing condition of 800 °C temperature, 1-hour holding time and 1-ton weight, the segmented YBCO-CCO sample showed two strongly connected layers with the relative density of about 96%. The X-ray diffraction (XRD) patterns indicated that each segment showed pure phase corresponding to each respective composition. Scanning electron microscopy (SEM) results confirmed the sharp interface and good adhesion between YBCO and CCO layers. Although the chemical analysis indicated the limited inter-layer diffusion near the interface, some elemental diffusion at the boundary was expected to be the source of this strong bonding.

1. Introduction

Nowadays, thermoelectric materials play a significant role in several devices such as calorimeter, laser diode sensor, thermostat calibrating bath and integrated circuit cooler [1] due to their ability in converting heat energy to electrical energy and vice versa. Due to the stability against high temperature and normal air atmosphere [2], thermoelectric oxide compounds have become one of the attractive materials for waste-heat recovery applications. In addition to the environmental stability, these oxides also possess several other advantages such as non-toxicity, lighter elements and low cost. Among oxide thermoelectrics, yttrium barium copper oxide ($YBa_2Cu_3O_{7-x}$ or YBCO) and calcium cobalt oxide ($Ca_3Co_4O_9$ or CCO) have been investigated for p-type thermoelectric leg. YBCO has an orthorhombic structure with triple oxygen-deficient perovskite layers. Its crystal structure and electrical conduction behavior could be changed with oxygen content [3]. The YBCO compounds showed a promising thermoelectric efficiency (i.e. ZT ~0.25-1.8) at low temperature [4, 5]. In case of CCO, the material has been intensively studied as a potential material for clean heat-to-electricity conversion at high temperature with the reported ZT value of ~0.3 [6, 7, 8].

Thermoelectric performance can be enhanced by employing a segmented (or layered) structure. In this case, two thermoelectric materials having maximum ZT at different temperature ranges are joined together to form bi- or tri-layers. In previous work on segmented oxide thermoelectrics, the conversion efficiency was found to increase by about 5% via a thermoelectric oxide couple between segmented p-HH/$Ca_3Co_4O_9$ and n-$Zn_{0.98}Al_{0.02}O$ [2]. To date, the segmented thermoelectric of YBCO and CCO has not yet been investigated. Thus, we are interested in fabricating the bi-layer of these materials using the hot pressing technique. Physical and chemical properties of this segmented YBCO-CCO ceramic such as phase, microstructure and chemical composition were also examined and discussed.

2. Materials and Methods

The $YBa_2Cu_3O_{7-x}$ (YBCO) and $Ca_3Co_4O_9$ (CCO) powders were synthesized by mixed oxide method using Y_2O_3 (Aldrich, 99.9%), $BaCO_3$ (Sigma-Aldrich, ≥ 99%), CuO (Sigma-Aldrich, 98%), $CaCO_3$ (Aldrich, 99.5%) and Co_3O_4 (Aldrich, 99.5%) as starting materials. The stoichiometric mixture of each compound was ball-milled in ethanol for 24 h and dried in an oven. These mixtures were then calcined at 850 °C for 12 h and at 750 °C for 24 h to form YBCO and CCO powders, respectively.

For the hot pressing process, YBCO powder was added first into a graphite mold. Then, CCO powder was added on top of the YBCO. The preparation steps are schematically shown in Figure 1(a). The segmented ceramics were fabricated using 1 ton of weight at 800 °C and 1 h holding time under Ar atmosphere. The phase formation and crystal structure of each layer was characterized by an X-ray diffractrometer (XRD, Lab X-6000) (see Figure 1 (b)). The relative density of the bilayer was calculated using Equation 1, where ρ is the density and V the volume fraction of materials [9],

$$\rho_{total} = \rho_{YBCO}V_{YBCO} + \rho_{CCO}V_{CCO} \quad (1)$$

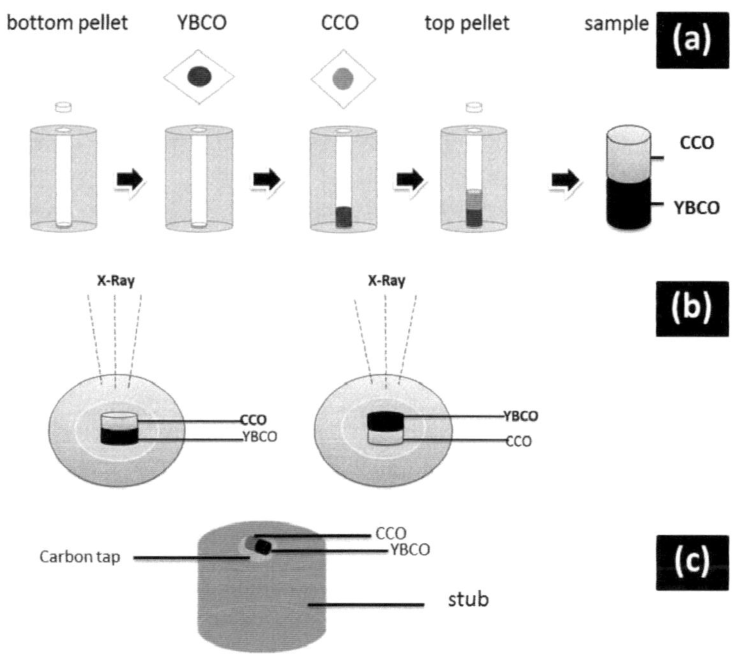

Figure 1. Schematic drawings of (a) steps of forming segmented YBCO-CCO ceramic, (b) XRD analysis of YBCO-side and CCO-side and (c) sample preparation for SEM/EDS analyses.

Scanning electron microscopy (SEM, JSM-IT300LV) was used for microstructural observation. The chemical composition was characterized by energy dispersive spectroscopy mode (EDS) of the SEM. Sample preparation for SEM and EDS involved cross-sectional grinding and polishing of the bilayer, ultrasonic cleaning and using the carbon-taped stub as the sample holder (Figure 1(c)).

3. Results and Discussion

The XRD patterns of calcined YBCO and CCO powders are shown in Figure 2 (a). YBCO was found to match with the JCPDS file no. 38-1433 ($YBa_2Cu_3O_7$) and CCO with JCPDS file no. 023-0110 ($Ca_3Co_4O_9$). Therefore, in case of YBCO, it was found that the material possessed the orthorhombic structure. This observation was in agreement with previous report [10]. In case of CCO, the material showed monoclinic structure. The calcined powders were virtually single phase. Other secondary phases were either not detected by XRD method or their intensities were below the noise level of XRD patterns. This suggested that the synthesis conditions were suitable for a complete reaction of the starting oxide precursors.

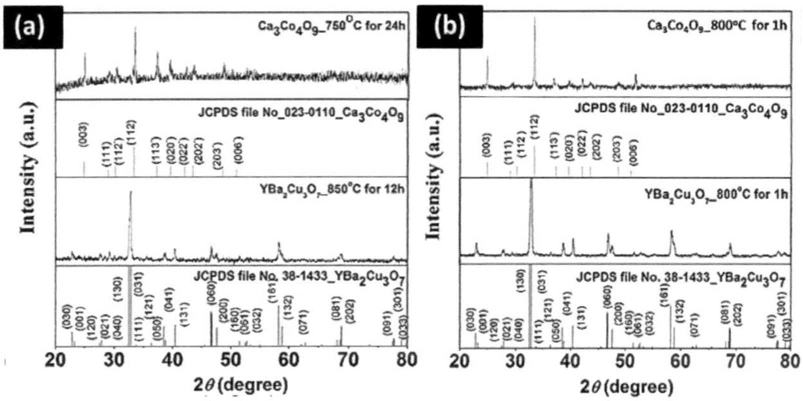

Figure 2. (a) XRD patterns of YBCO and CCO powders and (b) XRD patterns of segmented YBCO-CCO ceramic.

The XRD patterns of each side of the segmented YBCO-CCO ceramic are shown in Figure 2 (b). The phase of each compound was the same despite the ceramic being fabricated under Ar atmosphere. This indicated that the hot pressing condition used in this study was suitable for fabrication of YBCO and CCO ceramics. The XRD peaks in both YBCO and CCO compounds showed much higher intensity relative to the noise level. This was due to the larger grain sizes of the ceramics (i.e. length ~4.6-35.8 μm and width ~2.3-16.1 μm for plate-shaped YBCO and average size ~5 μm for CCO) compared to the particle sizes for the powder cases (i.e. ~1.7 μm for YBCO and ~0.2 μm for CCO).

SEM back-scattered electron images of segmented YBCO-CCO ceramics are presented in Figure 3 (a). In the figure, the boundary between YBCO and CCO ceramics could be seen clearly. The absence of cracks near the interface suggested that the strain caused by the difference in thermal expansion of these two compounds did not significantly affect the bonding process. Beside the observed connection at the interface, the YBCO ceramic showed bright color while the CCO ceramic showed dark color due to the fact that the former had a higher average atomic number (i.e. higher average number of electrons per atom and therefore higher electron scattering capability) than the latter. This phenomenon was typically used for discriminating the areas having different compositions such as secondary phases present in the matrix. However, in this study, there seemed to be only YBCO and CCO on each side of the bi-layered ceramic. The chemical compositions of some selected areas (see Figure 3 (a)) of the segmented YBCO-CCO ceramic are shown in Figure 3 (b). For point 1, 2, and 3, the element Y, Ba, Cu and O were detected. This confirmed that the bright area was YBCO material. On the other hand, point 5, 6, and 7 indicated a presence of Ca, Co and O elements indicating the CCO compound. Regarding the region near the interface between the two layers (point 4), the element Ba, Cu, Ca, Co and O were detected. This was attributed to the ionic diffusion and substitution. The tolerance factor was expected to be main reason for the substitution. In general, the ions having similar ionic radii can substitute one another. For the YBCO-CCO system, there seemed to be some compositional variation on the YBCO side as a function of position. For the position closer to the interface, Ba^{2+} showed an increasing trend while Y^{3+} and Cu^{2+} showed decreasing values. Although it was expected that some elemental diffusion and/or phase change occurred in these regions, further investigation was needed before any conclusion could be made. In contrast, the CCO side showed the elemental concentration to be more or less independent of position. This may suggest that CCO was a more stable compound compared to YBCO. Nevertheless, this bilayer structure showed that hot pressing is a viable method to produce high quality segmented oxide thermoelectric composing of two structurally and chemically different compounds.

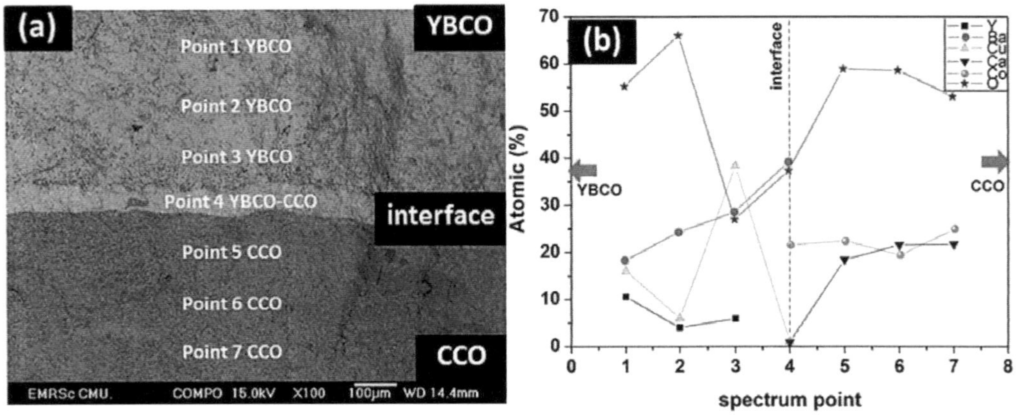

Figure 3. Microstructural investigation showing (a) SEM back-scattered image and (b) EDS analysis of the segmented YBCO-CCO ceramic.

4. Summary

$YBa_2Cu_3O_{7-x}$-$Ca_3Co_4O_9$ segmented ceramic with high relative density was successfully prepared by hot pressing method at the temperature of 800 °C for 1 h using the applied weight of 1 ton under an Ar atmosphere. The XRD pattern of each segment showed single phase without detected impurities. The SEM back-scattered image near the interface showed good bonding between the two layers. The result of chemical composition measurement by EDS mode suggested that the CCO side was rather chemically stable while the YBCO side showed a variation of the elemental components with respect to the distance from the interface. Despite the fact that there may be reaction and/or ionic diffusion at the interface, hot pressing seemed to be a suitable method to produce high-quality segmented thermoelectric ceramic for YBCO-CCO system.

5. Acknowledgements

This research was financially supported by the Thailand Research Fund (RSA5880005 and IRG5780013) and Thailand Institute of Scientific and Technological Research. Partial supports from the Center of Excellence in Materials Science and Technology, the National Research University Project under Thailand's Office of the Higher Education, Faculty of Science, and the Graduate School, Chiang Mai University are also acknowledged. We would also like to thank Center of Excellence in Alternative Energy, Sakon Nakhon Rajabhat University for the hot-press equipment. P. Wannasut would also like to thank the financial support from the TRF through the Royal Golden Jubilee Ph.D Program (PhD 0173/2558).

References

[1] Information on https://thermal.ferrotec.com/technology/thermoelectric-reference-guide/thermalref/ (available 8/6/2017)

[2] Le T H, Pryds N, Van N and Linderoth S 2014 Segmented Thermoelectric Oxide Based Module (Denmark: Technical University of Denmark)

[3] Howe B A 2014 Crystal Structure and Superconductivity of $YBa_2Cu_3O_{7-x}$, (Minnesota: Minnesota State University)

[4] Rodriguez J E 2005 *Physica Status Solidi* **2** **[10]** 3605-3608

[5] Prayoonphokkharat P and Watcharapasorn A 2016 *Science of Advanced Materials* **8** 1-4

[6] Xiang P H, Kinemuchi Y, Kaga H and Watari K 2008 *Journal of Alloys and Compounds* **454** 364-369

[7] Li N, Jing Y, Li G, Wang C, Shi J and Yu D 2009 *Journal of Alloys and Compounds* **467** 444-449

[8] Information on http://shodhganga.inflibnet.ac.in/bitstream/10603/37777/2/011_chapter%202.pdf (available 12/7/2017)

[9] Jaita P 2009 Effects of Bismuth Sodium Titanate and Doped Bismuth Sodium Titanate Additions on Structure and Properties of Lead Zirconate Titanate Ceramics, (Thailand: Chiang Mai University)

[10] Wannasut P, Keawprak N and Watcharapasorn A 2017 *Chiang Mai Journal of Science* **44(X)** 1-6.

IOP Conf. Series: Materials Science and Engineering **303** (2017) 012011 doi:10.1088/1757-899X/303/1/012011

Investigation on Thermal Properties of Kenaf Fibre Reinforced Polyurethane Bio-Composites

Mathan Athmalingam[1], W V Vicki[1]

[1]Centre for Advance Material (CAM), College of Engineering, Universiti Tenaga Nasional, Malaysia

vignesh@uniten.edu.my

Abstract. This research focuses on the effect of Kenaf fibre on thermal properties of Polyurethane (PU) reinforced kenaf bio-composites. The samples were prepared using the polymer casting method with different percentages of kenaf fibre content (5 wt%, 10 wt%, 15 wt%). The thermal properties of Kenaf/PU bio-composite are determined through the Thermogravimetric Analysis and Differential Scanning Calorimeter test. The TGA results revealed that 10 wt% Kenaf/PU bio-composite appeared to be more stable. DSC results show that the glass transition temperature (T_g) value of 10 wt% Kenaf/PU composite is significant to pure polyurethane. It can be said that the thermal stability of 10 wt% Kenaf/PU bio-composite exhibits higher thermal stability compared to other samples.

1.Introduction

Bio-composites are the most suitable materials significant in nature for their utilization in different fields due to their eco-friendly features. Bio-composites have many advantages; they are relatively cost effective, display great thermal and stability and have low coefficient of friction and low density [1]. The utilization of sustainable sources like natural fibres in making thermal insulation board is expanding quickly in numerous enterprises. People now are venturing further to discover ways to save the earth by developing products that have high biodegradability and at the same time low energy consumption [2]. Natural fibres have a high specific heat-storage capacity and can be utilized to reduce overheating in buildings. The high density and complex microstructure of natural fibre insulations can likewise give great acoustic insulation. Natural fibre insulations are non-toxic. The utilization of natural fibres are developing in numerous areas, for example, in vehicles, furniture, and construction. This is for the most part because of their advantages in contrast with synthetic fibres, i.e. low cost, low weight, less damage to processing equipment, great relative mechanical properties, bottomless and sustainable resources [3-4]. The use of bio-composites from renewable resources has increased significantly, generally, due to their biodegradable nature. Bio-composites have many advantages; they are relatively cost effective, display great thermal stability, low coefficient of friction and low density. Bio-composites are principally utilized as part of top-line applications, for example, automotive industries, aeronautical engineering, and so on [5]. One of the problems that have been discovered from existing insulation applications is with insulation using synthetic material (e.g polyurethane) which is not very eco-friendly. It is expensive and non-biodegradable. This study was conducted to develop an eco-friendly thermal insulation material using natural fibres. This was achieved by decreasing the use of synthetic based filler materials and increasing the usage of natural fibres. Polyurethanes are a standout amongst the most adaptable materials on the planet today. It has been connected to a wide range of ventures, for example, adaptable foam in upholstered furniture, as an inflexible foam insulation in walls, ceilings, and appliances of thermoplastic produced using polyurethane in medical devices and footwear, adhesives, sealants and plasters utilized on floors and car insides [6]. Kenaf fibres are acquired from the Hibiscus cannabinus, a quick growing plant ready to achieve 3.5 m of height in 2 years. The kenaf products are environment friendly and generally used in developing window frames, interior wall panels, kitchen cupboards and as a floor substitute in replacement of parquet and cement floors [7].

2.Materials

Polyurethane is a combination of both the resin and the hardener. Mirathane 6414 A/B are used for this project. The resin is known as Mirathane 6414 A and it is beige in colour with a viscosity of 2000-4000 CPS at 25^{0}C. The hardener is known as Mirathane 6414 B and it is brown in colour with a viscosity of 30-80 CPS at 25^{0}C. The mixing ratio of both this resin and hardener is 100:64. The Polyurethane that was used in this study was supplied by Miracon (M) Sdn. Bhd, Malaysia. Kenaf short fibre of size < 212 μm with tensile strength, modulus and elongation are 930MPa, 53GPa, and 1.6% respectively. The kenaf short fibre was purchased from Innovative Pultrusion Sdn Bhd.

3. Bio-composite Preparation

Polyurethane and kenaf composition were prepared using the polymer casting method. Three different fiber loading composition were prepared with 5, 10 and 15 %wt of kenaf fibre. The desired weight in grams of the Polyurethane resin, hardener and Kenaf fibre were measured using a digital weighing scale and mixed. The mixture was then stirred thoroughly to obtain uniform mixture of polyurethane anc kenaf fibre. Then, the mixture was poured into mould and left to cure for 24 hours at room temperature. Hardened samples were removed from the mould and proceeded for testing

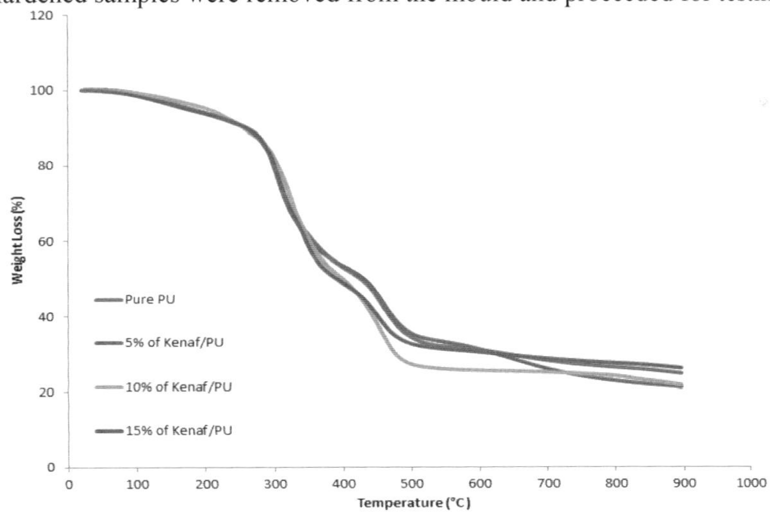

Figure 1. TGA curves for Kenaf/PU bio-composite with different fibre loading

4. Results and discussion

4.1 Thermogravimetric Analysis (TGA).

Thermogravimetric analysis was done to measure changes in the weight reduction (mass) of samples that were subjected to a steady increment of temperature to evaluate reactions including gaseous emissions. At temperatures lower than 100°C, the samples lost weight by undergoing water evaporation. Table 1 shows the decomposition temperature range, residue percentage and weight loss percentage for kenaf/PU bio-composite with different fibre loading. The TGA and DTG curves of pure PU, and Kenaf/PU bio-composite with 5, 10, and 15 wt% fibre loading are represented in figure 1 and figure 2.

The initial mass loss for every composite occurred approximately around 200 – 400°C with a maximum loss at 300°C corresponding to a weight loss of about 50%. 10 wt% Kenaf/PU bio-composite had an initial mass loss at 253.33°C which is much earlier than PU, 5 wt% Kenaf/PU and 15 wt% Kenaf/PU. This is attributed to the higher thermal stability of fibre and phase compatibility between polyurethane and kenaf. The reason for the mass loss is because there is a deterioration of the hemicellulose from kenaf and also a disturbance in the urethane bond. When temperature increases, more raw materials vaporize or decompose, so mass continues to reduce. This denoted a reduction in thermal stability from the addition of kenaf fibre into the matrix. Norshahida Sarifuddin et. al (2013) [8] concluded that it is best to say the reduction in weight loss was the result of improvement in thermal stability [8]. It is also said from the previous studies conducted by Sunny M. Ogbomo et. al 2013 [9], natural fibres with hemicellulose tend to have low thermal stability.

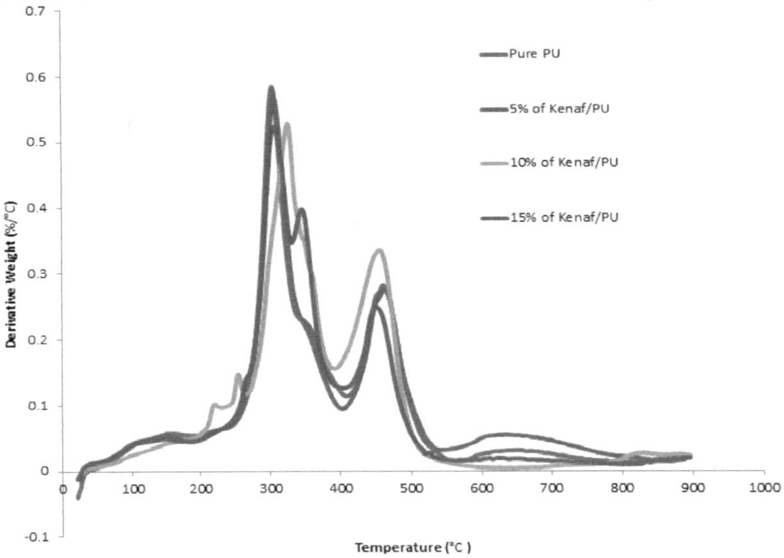

Figure 2. Differential Thermal Analysis (DTA) curves

This was due to the early decomposition of kenaf in the composition while the pure PU decomposes awhile later compared to kenaf/PU composites. The second decomposition occurred in a range of 350–500 °C with the main peak picturized at around 450 °C with a weight loss of about 70%. This two-step degradation process demonstrated that the incorporation of Kenaf fibre did not affect the decomposition behavior of pure PU. The percentage of weight at 900°C reflects the amount of residue left after the samples were completely degraded. Pure PU and 15% wt Kenaf/PU composite has the highest residue percentages of 24.74% and 26.13% than 5% and 10% wt Kenaf/PU composite. It can be said from previous studies conducted by Y.A. El-Shekeil et al 2012 [10] that the lowest weight loss percentage has the highest thermal stability.

Table 1. TGA results of kenaf/PU bio-composite with different fibre loading.

Sample	Decomposition temperature range (°C)	DTGmax (°C)	Maximum Decomposition	
			Residue (%)	Weight loss (%)
Pure PU	270.40 - 544.00	303.42	24.74	75.26
5 wt% Kenaf	267.29 - 584.88	306.08	21.13	78.87

10 wt% Kenaf	253.33 -7 77.22	325.20	21.71	78.29
15 %wt Kenaf	242.42 - 618.50	306.08	26.13	73.87

DTG curves of PU/kenaf composites is illustrated in figure 2. Maximum weight loss occurred at temperature Tmax and was determined from the DTG curves. It can be seen that the main peaks of kenaf composites have lower values compared to pure polyurethane composites. From these results, it is clear that the thermal stability and decomposition temperature of the composites are affected by the percentage of kenaf composition in polyurethane. Good interfacial adhesion between the kenaf fibre and polyurethane would result in good compatibility between fibre and matrix where it shows an increase in degradation temperature of composites compared to pure PU.

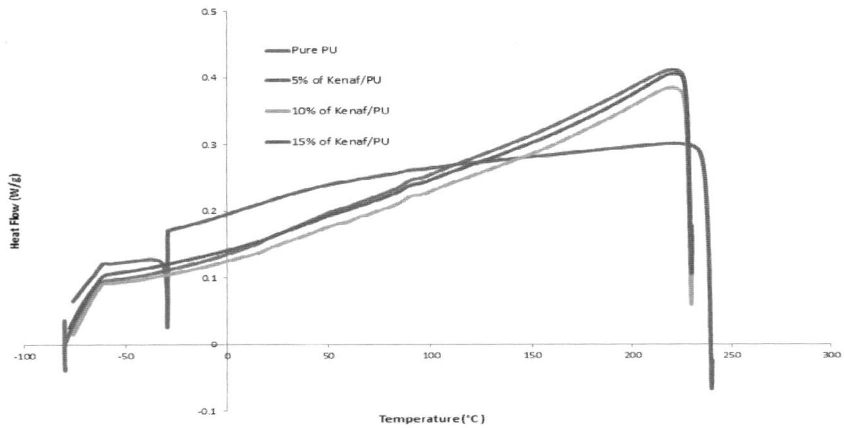

Figure 3. DSC thermograms (heating-step)

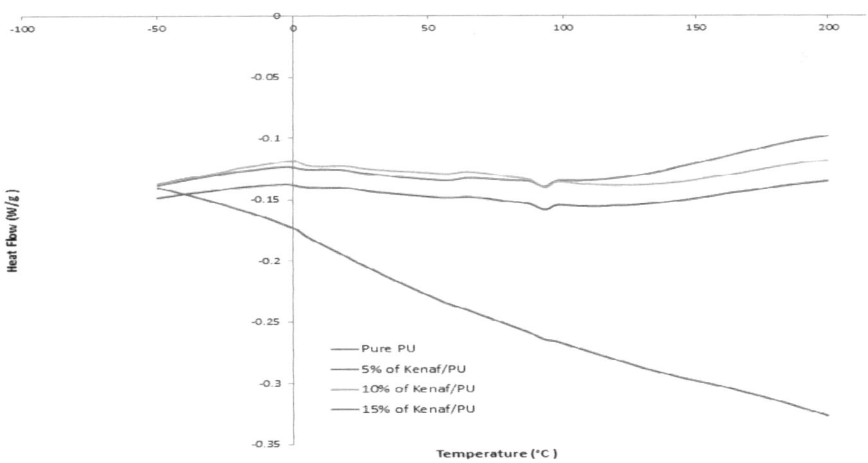

Figure 4. DSC thermograms (cooling-step)

4.2 Differential Scanning Calorimeter (DSC) analysis

DSC analysis gives the flow rate associated with a thermal event as a function of time and temperature to obtain quantitative information about melting and phase transition of polymeric materials. As can be seen in the figure 3 and figure 4, the glass transition temperature (T_g) for pure polyurethane occurred at 95.27°C, and there was no significant change in Tg for other composites. The cold crystallization peak temperature appears at 90 °C. A decrease in the cold crystallization temperature indicates enhancement in crystallinity. The T_m of the crystalline component in a blend is dependent on both morphological and thermodynamic factors. The crystallinity of 5 wt% Kenaf/PU bio-composite was higher with lower fibre content than the 10 wt% and 15 wt% composite, indicating that the presence of a small amount of fibre increases the crystallization of the composite.

5. Conclusion

From the present study, it can be concluded that kenaf fibre reinforced polyurethane bio-composite was successfully prepared. Based on analysis, the thermal test results have shown good positive trends. TGA and DTG curves revealed that (10 wt% Kenaf/PU) appeared to be more stable because the higher DTG max is noticed from the resulted graph. DSC results show that the Tg value of (10 wt% Kenaf/PU) is similar to polyurethane. It can be said that the thermal stability neither improved nor deteriorated. Therefore, the addition of natural fibres does not have significant negative effects to the PU bio-composite and this can be a strong evidence to develop bio-composites with high thermal stability.

6. Acknowledgement

The authors would like to acknowledge the financial support provided by Universiti Tenaga Nasional under the BOLD 2025 research grant scheme to promote research excellence that supports both nation's and TNB needs.

7. References

[1] Keynakli O 2012 *Renewable and Sustainable Energy Reviews* **16** 415–425
[2] Korjenic A, Zach J, and Hroudová J 2016 *Energy and Buildings* **116** 45-58
[3] Sanjay M R., Arpitha G R, Naik L L, Gopalakrishna K and Yogesha B *Natural Resources* ***07*(03)** 108-114
[4] Mohamed Amine Laadila, Krishnamoorthy Hegde, Tarek Rouissi, Satinder Kaur Brar, Rosa Galvez, Luca Sorelli, Ridha Ben Cheikh, Maria Paiva, Kofi Abokitse 2017 Journal of *Cleaner Production* **164** 575-586
[5] P Asokan, M Firdoous and W Sonal 2012. *Rev.Adv.Mater. Sci.* **30** 254-261
[6] Mawarnie Ismail, Farhana Mohd and Abdul Rahim Bahari 2016 *Journal of Engineering and Applied Sciences* **11** No. 16
[7] Edynoor bin Osman, Toshihiro Moriga, Kei-ichiro Murai, Mohd Warikh bin Abd Rashid 2017 *Industrial Crops and Products* **100** 171-175
[8] Sarifuddin, N, Ismail H, and Ahmad Z 2013 *Journal of Thermoplastic Composite Materials* ***28*(4)** 445-460
[9] Ogbomo S M, Ayre B, Webber C L, and D'Souza N A. *Polymer Composites*, ***35*(5)** 915- 924
[10] El-Shekeil Y, Sapuan S, Zainudin E, and Khalina A 2011 **471** 58–63

Grain refinement of Al-$Si_{9.8}$-$Cu_{3.4}$ alloy by novel Al-3.5FeNb-1.5C master alloy and its effect on mechanical properties

K Ch Apparao[1] and Anil Kumar Birru[*2]

[1, 2] Department of Mechanical Engineering,
[1, 2] National Institute of Technology, Manipur- India-795004

anilbirru@gmail.com

Abstract. A novel Al-3.5FeNb-1.5C master alloy with uniform microstructure was prepared using a melt reaction process for this study. In the master alloy, basic intermetallic particles such as $NbAl_3$, NbC act as heterogeneous nucleation substrates during the solidification of aluminium. The grain refining performance of the novel master alloy on Al-$Si_{9.8}$-$Cu_{3.4}$ alloy has also been investigated. It is observed that the addition of 0.1 wt.% of Al-3.5FeNb-1.5C master alloy can induce very effective grain refinement of the Al-$Si_{9.8}$-$Cu_{3.4}$ alloy. The average grain size of α-Al is reduced to 22.90 μm from about 61.22 μm and most importantly, the inoculation of Al-$Si_{9.8}$-$Cu_{3.4}$ alloy with FeNb-C is not characterised by any visible poisoning effect, which is the drawback of using commercial Al-Ti-B master alloys on aluminium cast alloys. Therefore, the mechanical properties of the Al-$Si_{9.8}$-$Cu_{3.4}$ alloy have been improved obviously by the addition of the 0.1 wt.% of Al-3.5FeNb-1.5C master alloy, including the yield strength and elongation.

1. Introduction

Some of the most important driving forces for the development of Al-Si cast alloys are the high strength to weight ratio, high fluidity, low shrinkage in casting, superior wear resistance, low coefficient of thermal expansion (CTE), high corrosion resistance and excellent castability which makes them potential materials for a number of applications in automobiles and other engineering sectors [1-3]. It is well known that the mechanical properties of aluminium alloys can be improved by means of grain refinement. Actually, this is quite well-established industrial practice for wrought aluminium alloys, where the refinement is done by means of Al-Ti-B master alloys [4-7]. The Al-Ti-B master alloy offers a remarkable performance in the continuous and semi-continuous casting of wrought alloys but fails to meet the expectations in the case of aluminium cast alloys, especially for the Al-Si alloys with a content of Si larger than 4% [8-10]. The reason for this drawback is that the Si in the melt will react with Ti to form titanium silicides which prevent the effective grain refinement of the alloy, a phenomenon known as poisoning [11-12]. By now, various attempts have been done to solve this problem [13-15]. One effective method is to reduce the Ti content in the mater alloys. Some new master alloys, such as Al-3Ti-3B, Al-1Ti-3B, Al-3B and Al-3Ti-1B-0.2C have been reported to provide better grain refinement performances than the Al-5Ti-1B master alloy [16-17]. However, these methods have not been adopted and grain refinement problems of Al-Si alloys are still unresolved in Al cast industry. To overcome this poisoning effect, recently Nb-B grain refiner can efficiently refine binary Al-xSi alloys due to the formation of niobium borides which are more stable than titanium

IOP Conf. Series: Materials Science and Engineering **303** (2017) 012012 doi:10.1088/1757-899X/303/1/012012

borides, therefore, appreciably limiting the so called poisoning effect [18]. From an industrial point of view, availability and cost of material are very important. Due to adequate availability and less cast of Ferro niobium (FeNb) and Carbon powder (C), in this study Al-FeNb- C has been taken as grain refiner for grain refinement of α-Al of the commercial Al-$Si_{9.8}$-$Cu_{3.4}$ alloy by means of heterogeneous nucleation.

2. Materials and Methods

The raw materials employed in this study are: Pure Aluminium (99.5%), Ferroniobium metal (60% of Niobium), Carbon powder (99.5%) and Al-Si alloy (Al-$Si_{9.8}$-$Cu_{3.4}$). The required amounts of Aluminium, FerroNiobium and Carbon powder were prepared in order to fabricate an Al-3.5FeNb-1.5C master alloy. Aluminium was melted at 790 ^{0}C for 1hour and, subsequently, the carbon powder was added to the melt and the temperature was increased to 850 ^{0}C. After 1hour at 850 ^{0}C, Ferro-niobium metal was added and left to dissolve for 2 hours with intermediate stirring. The microstructure of the Al-3.5FeNb-1.5C master alloy was analysed using Scanning Electron Microscopy (SEM)

The present study was performed on Al-$Si_{9.8}$-$Cu_{3.4}$ alloy with a chemical composition given in Table 1. A standard 230-tons cold chamber High Pressure Die Casting (HPDC) machine was used to fabricate the cast samples at 720 ^{0}C molten temperature. Prior to casting, the melt was treated with different weight percentages of the grain refiner such as 0.1 and 1.0 wt.% of Al–3.5FeNb–1.5C master alloy were added to the mix and then poured into a pouring hole of the HPDC machine to perform casting. After the Die casting experimentation process, the casted part has been taken from the die. Once cast, the samples were cut and their cross-sections prepared for metallographic analysis by using the standardized route of grinding with 120-1200 SiC papers plus polishing with OPS solution was employed. Macroetched cross-section samples were etched by means of Tucker's solution (15ml HF + 15ml HNO_3 + 45ml HCl + 25ml H_2O). The microstructural analysis was carried out with a Carl Zeiss Axioscope A1 optical microscope. Also, the microstructure and the extracted particles of the Al-3.5FeNb-1.5C master alloy were analyzed by X-ray diffraction (XRD) and Field Emission Scanning Electron Microscope (FESEM).

Table 1.Chemical composition of Al-Si9.8-Cu3.4 alloy (wt.%).

Si	Cu	Fe	Ti	Pb	Sn	Al
9.839	3.474	0.189	0.184	0.043	0.013	Balance

3. Results and discussion

3.1 Microstructure and microanalysis of the Al-3.5FeNb-1.5C master alloy

Figure 1 shows the XRD results of the Al-3.5FeNb-1.5C master alloy. It has been seen that the master alloy mainly contains four kinds of phases: α-Al, $NbAl_3$, NbC and FeC. These intermetallic particles ($NbAl_3$ and NbC) act as a heterogeneous nucleation substrates during the solidification of Aluminium. Also NbC particles control the agglomeration of FeC particles. The microstructure of the Al-3.5FeNb-1.5C master alloy is shown in the Figure 2. It may be seen that the second phase particles are dispersed in the aluminium matrix homogeneously. As shown in Figure 2, the particles are not connected with each other and the size is various. Most of the particles are large such as $NbAl_3$, NbC surrounded by lots of smaller particles.

3.2 Effect of FeNb-C inoculation in the binary Al-$Si_{9.8}$-$Cu_{3.4}$ alloy.

The microstructures of the Al-$Si_{9.8}$-$Cu_{3.4}$ alloy before and after the addition of Al-3.5FeNb-1.5C master alloys are shown in Figure 3. It may be seen from Figure 3(a) that the grain structure of the unrefined binary Al-$Si_{9.8}$-$Cu_{3.4}$ alloy is coarse with an average grain size of approximately 62µm. These coarse grains are obviously refined to small equiaxed ones by adding the Al-3.5FeNb-1.5C master alloy.

Notably, when inoculated with 0.1wt.% of Al-3.5FeNb-1.5C master alloy, the grains of the Al-$Si_{9.8}$-$Cu_{3.4}$ alloys are very fine as shown in Figure 3(b) and uniform for the entire range of holding times, implying remarkable grain refining efficiency. After the addition of 0.1wt.% of Al-3.5FeNb-1.5C master alloy, the average grain size of α-Al is 22μm which is much finer than those refined by the 1.0wt.% of grain refiner. The microstructures of Al-$Si_{9.8}$-$Cu_{3.4}$ alloy with the addition of 1.0 wt. % of Al–3.5FeNb–1.5C grain refiner are shown in Figure 3(c). From this microstructure, it could be understood that the grain sizes are again increased because of the formation of β-Al_5FeSi platelets which are cause to the formation of coarse grains. These platelets (or needles as they appear in the microstructure) often result in the formation of large shrinkage cavities due to the inability of the liquid metal to fill the spaces between the branched platelets. From these results, it can be concluded that the Al-3.5FeNb-1.5C master alloy has a much better grain refining performance particularly at 0.1 wt. % of grain refiner on Al-$Si_{9.8}$-$Cu_{3.4}$ alloy; also it shows a strong anti-fading effect in Al-Si casting alloys.

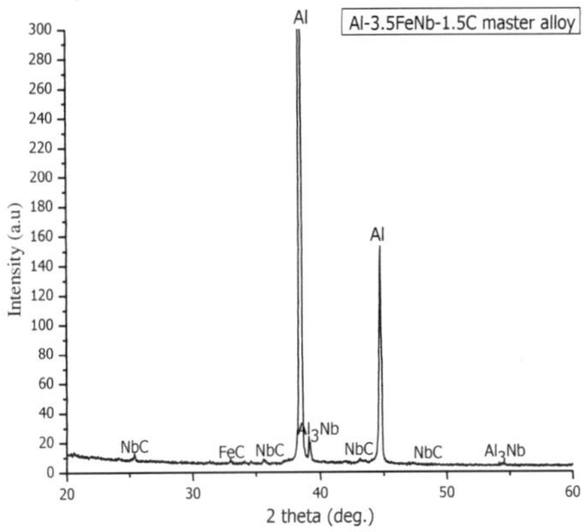

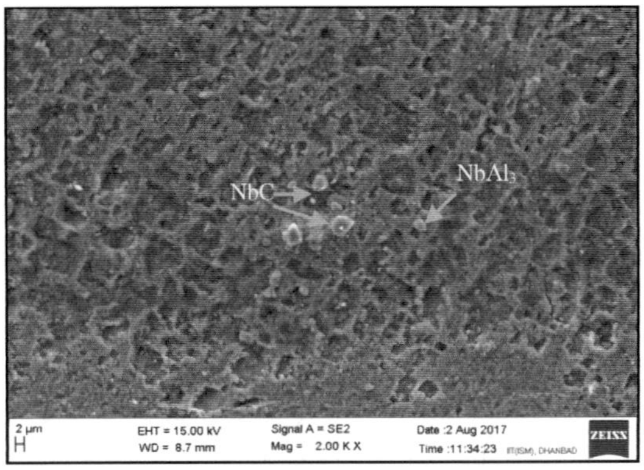

Figure 1. XRD results of the Al-3.5FeNb-1.5C. **Figure 2**. SEM microstructure of Al-3.5FeNb-1.5C.

3.3 Mechanical properties of the Al-$Si_{9.8}$-$Cu_{3.4}$ alloy

The grain refinement effect on the mechanical properties of Al-$Si_{9.8}$-$Cu_{3.4}$ alloy is investigated in this work, including the yield strength (YS) and elongation. The YS and elongation of the Al-$Si_{9.8}$-$Cu_{3.4}$ alloys before and after addition of grain refinement are shown in Table.2 and Figure 4. As expected, the refined microstructure could improve the tensile properties of Al-$Si_{9.8}$-$Cu_{3.4}$ alloy. Before the addition of grain refiner, the YS and elongation values are 186.706 MPa and 2.76% as shown in the Table 2. After applying 0.1 wt.% of Al-3.5FeNb-1.5C master alloy, the YS and elongation values are 204.73MPa and 4.48% for Al-$Si_{9.8}$-$Cu_{3.4}$ alloy. From these results, it was cleared that the YS and elongation of the Al-$Si_{9.8}$-$Cu_{3.4}$ alloys are significantly improved by the addition of grain refiner, particularly at 0.1 wt. % of Al-3.5FeNb-1.5C. After the addition of 0.1 wt. % of Al-3.5FeNb-1.5C master alloy, that is increased from 186 MPa to 204 MPa. That is to say that the average value of the YS is increased by 9.7%. It is also worth noticing that the refinement using 0.1wt.% of Al-3.5FeNb-1.5C could double the elongation of Al-$Si_{9.8}$-$Cu_{3.4}$ alloy than that of without refinement. But the YS and elongation of the Al-$Si_{9.8}$-$Cu_{3.4}$ alloys are decreased instead of improved by the addition of grain refiner, particularly at 1.0 wt. % of Al-3.5FeNb-1.5C, because of the formation of β-Al_5FeSi platelets which causes to formation of the shrinkage cavities due to the inability of the liquid metal to fill the spaces between the branched platelets.

IOP Conf. Series: Materials Science and Engineering **303** (2017) 012012 doi:10.1088/1757-899X/303/1/012012

Table 2. Grain size and mechanical property data of Al-$Si_{9.8}$-$Cu_{3.4}$ cast alloy

Wt.% of Grain refinement	Grain size(μm)	Mechanical property	
		Yield Strength (MPa)	Elongation (%)
Without grain refiner	61.22	186.706	2.76
0.1 wt.% of grain refiner	22.9	204.73	4.48
1.0 wt.% of grain refiner	43.07	191.874	3.36

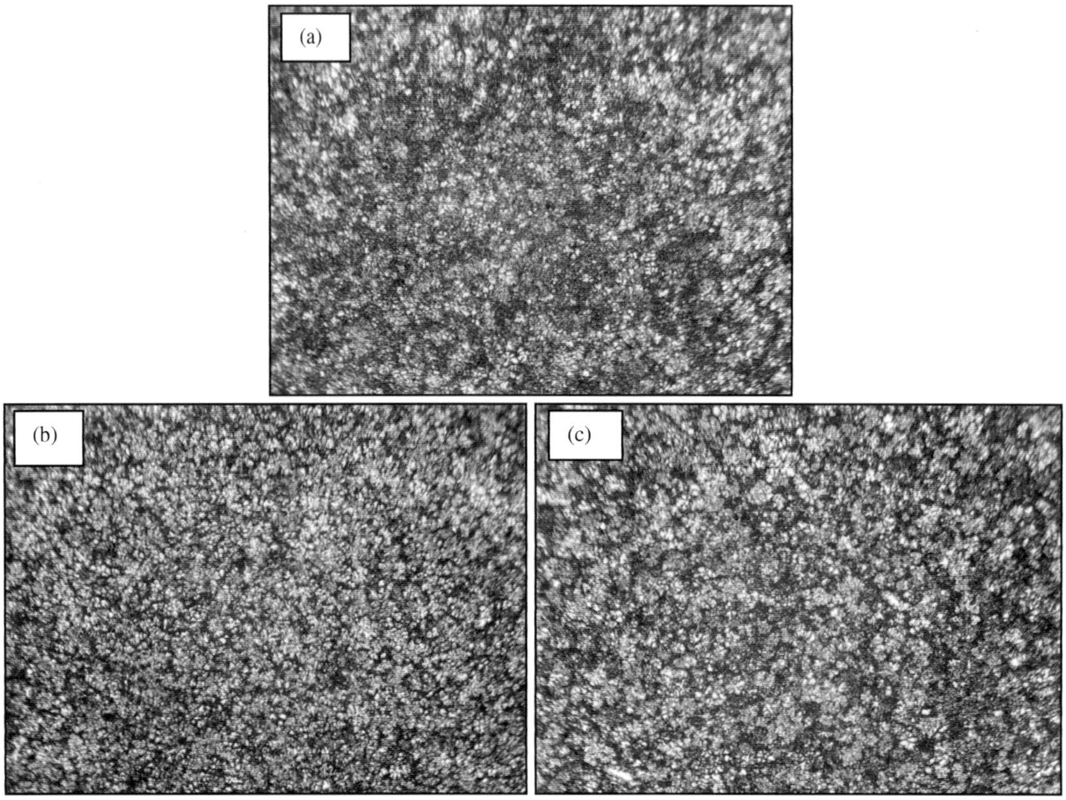

Figure 3. Optical microphotographs of as-cast Al-$Si_{9.8}$-$Cu_{3.4}$ alloy at 100X magnification: (a) un-treated (without grain refiner), (b) 0.1 wt. %: of Al–3.5FeNb–1.5C, (c) 0.3 wt. % of Al–3.5FeNb–1.5C.

Table 2 and Figure 4 also show the change of YS and elongation of the Al-$Si_{9.8}$-$Cu_{3.4}$ alloy after the addition of 1.0 wt. % of the Al-3.5FeNb-1.5C master alloy. Finally, it is worth noting that the sample with 0.1 wt. % of the Al-3.5FeNb-1.5C master alloy addition possesses the highest YS and % elongation. The result indicates that the grain refinement has resulted in a very evident improvement in the ductility of the Al-$Si_{9.8}$-$Cu_{3.4}$ alloy.

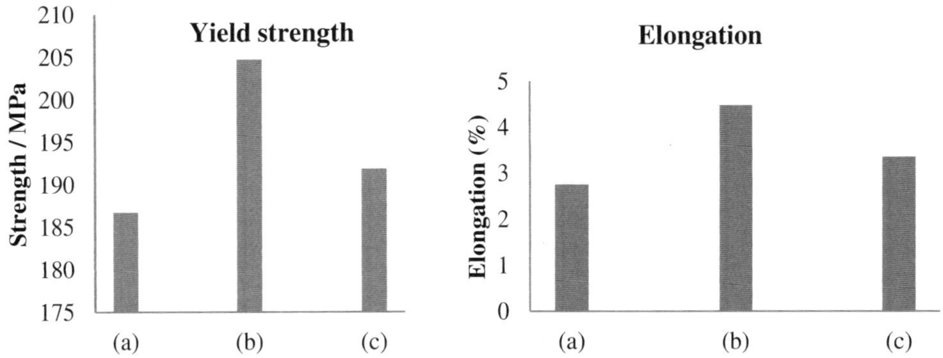

Figure 4. Yield strength and Elongation of Al-$Si_{9.8}$-$Cu_{3.4}$ at room temperature: (a) without addition of master alloy; (b) addition of 0.1 wt. % Al-3.5FeNb-1.5C; (c) addition of 1.0 wt. % Al-3.5FeNb-1.5C.

4. Conclusions

In the present work, a novel Al-3.5FeNb-1.5C master alloy with homogeneous microstructure was prepared and its effects on the microstructure and mechanical properties of Al-$Si_{9.8}$-$Cu_{3.4}$ alloy were studied. Based on the experimental results, the conclusions have been drawn as follows:

- In the novel Al-3.5FeNb-1.5C master alloy, basic intermetallic particles such as $NbAl_3$, NbC act as heterogeneous nucleation substrates during the solidification of Aluminium. Also, NbC particles control the agglomeration of FeC particles.

- Significant grain refinement performance of novel Al-3.5FeNb-1.5C master alloy in Al-$Si_{9.8}$-$Cu_{3.4}$ alloy was observed, particularly at 0.1 wt. % of Al-3.5FeNb-1.5C master alloy. The average grain size of the α-Al in the Al-$Si_{9.8}$-$Cu_{3.4}$ alloy is reduced to 22.90 μm from 61.22 μm by the addition of 0.1 wt. % of Al-3.5FeNb-1.5C master alloy.

- The mechanical properties of the Al-$Si_{9.8}$-$Cu_{3.4}$ alloy have been obviously improved by the addition of the 0.1 wt.% of Al-3.5FeNb-1.5C master alloy. The yield strength of the Al-$Si_{9.8}$-$Cu_{3.4}$ alloy is increased by 9.7%, as well as the average value of the percentage elongation is increased from 2.76% to 4.48%.

References

[1] Kori SA, Murthy BS and Chakraborthy M 2000 *Mater. Sci. Eng. A* **283** 94
[2] Lasa L and Ibade J 2002 *Scr. Mater.* **46** 477
[3] Sajjadi SA, Ezatpour HR and Torabi PM 2012 *Mater. Des.* 2012 **34** 106
[4] Han YF, Li K, Wang J, Shu D and Sun BD 2005 *Mater. Sci. Eng. A* **405** 306
[5] Tongmin W, Hong WF, Zongning C, Jing Z, Fei C and Tingju L 2012 *J. Alloys Comp.* **5 11** 45
[6] Vinod Kumar G, Murty BS and Chakraborty M 2009 *J. Alloys Comp.* **472** 112
[7] Schaffer PL and Dahle AK 2005 *Mater. Sci. Eng. A* **413** 373
[8] Zhu M, Jian ZY, Yang GC and Zhou YH 2012 *Mater. Des.* **36** 243
[9] Birol Y 2012 *J Alloys Compd.* **513** 150
[10] Qiu D, Taylor JA, Zhang M and Kelly PM 2007 *Acta Mater.* **55** 1447
[11] Shabani MJ, Emamy M and Nemati N 2011 *Mater. Des.* **32** 1542
[12] Kori SA, Auradi V, Murty BS and Chakraborty M 2005 *Mater. Forum* **29** 387
[13] Sigworth GK and Guzowski MM 1985 *AFS Trans.* **93** 907
[14] Kori SA, Murty BS and Chakraborty M 1999 *Mater. Sci. Technol.* **15** 986
[15] Birol Y 2007 *J Alloys Compd.* **443** 94

[16] Gezer BT Toptan F Daglilar S and Kerti I 2010 *Mater. Des.* **31** 530
[17] Li P, Liu S, Zhang L and Liu X 2013 *Mater. Des.* **47** 522
[18] Nowak M, Bolzoni L and Hari Babu N 2015 *Mater. Des.* **66** 366

Acknowledgement

Authors would like to acknowledge for the Financial Support of the National Institute of Technology (NIT), Manipur, Imphal-India to carry out the experimental research work of Mr. K. Ch. Apparao, Research Scholar (Enrolment Number 15PME001) under the guidance of Dr. Anil Kumar Birru, Assistant professor and Head in Mechanical Engineering Department- India-795004.

IOP Conf. Series: Materials Science and Engineering **303** (2017) 012013 doi:10.1088/1757-899X/303/1/012013

Monitoring of Surface Roughness in Aluminium Turning Process

Atitaya Chaijareenont[1] and Somkiat Tangjitsitcharoen[2,a*]

[1,2] Department of Industrial Engineering, Faculty of Engineering, Chulalongkorn University, Bangkok, 10330, Thailand

[a*]Somkiat.ta@eng.chula.ac.th

Abstract. As the turning process is one of the most necessary process. The surface roughness has been considered for the quality of workpiece. There are many factors which affect the surface roughness. Hence, the objective of this research is to monitor the relation between the surface roughness and the cutting forces in aluminium turning process with a wide range of cutting conditions. The coated carbide tool and aluminium alloy (Al 6063) are used for this experiment. The cutting parameters are investigated to analyze the effects of them on the surface roughness which are the cutting speed, the feed rate, the tool nose radius and the depth of cut. In the case of this research, the dynamometer is installed in the turret of CNC turning machine to generate a signal while turning. The relation between dynamic cutting forces and the surface roughness profile is examined by applying the Fast Fourier Transform (FFT). The experimentally obtained results showed that the cutting force depends on the cutting condition. The surface roughness can be improved when increasing the cutting speed and the tool nose radius in contrast to the feed rate and the depth of cut. The relation between the cutting parameters and the surface roughness can be explained by the in-process cutting forces. It is understood that the in-process cutting forces are able to predict the surface roughness in the further research.

1. Introduction

A turning process is widely used to produce the parts in industry such as the automotive parts. It is known that the CNC turning machine can manufacture the complicated mechanical parts in large volume and high accuracy, which can reduce cost and production time, including the high efficiency production rate. Normally, the surface roughness is an important indicator to decide the workpiece quality before sending to the customer. Owing to, the surface finish of the workpiece impacts the assembly process, especially the shaft of the spindle motor parts that have high precision. In fact, there are many factors affecting the surface roughness. However, surface roughness cannot be measured during the turning process.

However, the relation among the surface roughness, the cutting conditions and the cutting force in CNC turning process had been studied so far. There are many researches that have analyzed and investigated the surface roughness in the turning process. Thamizhmanii, Saparudin, and Hasan [1] analyzed of optimum cutting conditions to find the lowest surface roughness in turning SCM 440 alloy steel. The experimental results showed that the depth of cut and the feed rate affected the surface roughness. Rao, C.J., D.N. Rao and Srihari [2] studied the significance of cutting speed, feed rate and depth of cut on cutting force and surface roughness in the turning process. It has been proved that the feed rate has a significant influence on both the cutting force and the surface roughness. The depth of cut has influenced on the cutting force. Lalwani, Mehta, and Jain [3] investigated the effect of cutting parameters on feed force, thrust force, cutting force and surface roughness in finish hard turning of MDN250 steel. It was known that the effect of feed rate is the most significant factor on the surface roughness. The feed rate and the depth of cut affected cutting force and feed force.

The surface roughness models have been proposed by many researchers [4-7]. Tangjitsitcharoen [7] developed the in-process surface roughness model that uses the cutting force ratio by employing the exponential function. It has been proved that the cutting force can be used to predict the in-process surface roughness under various cutting conditions during the cutting.

The previous researches had focused on the parameters of carbon steel turning process and predicted the surface roughness using the cutting force ratio which consists of the main force and the feed force [6, 7]. The quality of surface roughness subjected to the feed rate, the cutting speed, the depth of cut and the tool nose radius [7]. Hence, the aim of this research is to study the relations among the surface roughness, the cutting conditions and the cutting forces during the turning process using the coated carbide tool and the work material of aluminium (Al 6063).

2. Relations of Cutting Condition, Cutting Forces and Surface Roughness

The important cutting conditions relate to the surface finish in CNC turning process are the feed rate, the cutting speed, the depth of cut and the tool nose radius. As the theory of surface roughness, the feed rate and the tool nose radius are directly concerned in the surface finish. The larger tool nose radius results in the low theoretical surface roughness while the higher feed rate causes the high theoretical surface roughness *[*8, 9*]*.

The high cutting speed leads to an increase in the cutting temperature so the work material becomes soft. It means that the cutting forces are lower and easy to cut, which results in a good quality of surface finish *[*7, 10, 11*]*. The large depth of cut brings about the larger cutting area and the higher dynamic cutting force which affects the vibration of cutting tool and workpiece during the cutting. Furthermore, it leads to an increase in the surface roughness *[7]*. Consequently, the relationship between the surface roughness and the above cutting parameters is considered due to the improvement of the quality of surface finish.

3. Experimental Setup and Procedure

The experiments are tested on the CNC turning machine. The aluminium alloy (Al 6063) is adopted for this experiment with the coated carbide tool, which is the different tool nose radiuses of 0.4 and 0.8 mm. The cutting tool's rake angle is 11°. The dynamometer (Kistler, model: 9121) is installed in the turret to generate a signal while turning. The cutting forces are amplified through the charge amplifier (Kistler, model: 5083) and demonstrate the result by the oscilloscope (Yokogawa, DL750) before digitization and calculation in the computer as shown in figure 1. The arithmetic mean surface roughness (Ra) and the surface roughness depth (Rz) are measured by the surface roughness tester (Mitutoyo, SJ-400). The cutting conditions are summarized in table 1.

The experimental setup and procedures are presented as follows:

1. To prepare the dimensional cylindrical workpieces.
2. To set the oscilloscope (at the sampling rate of 1kHz and the low-pass filtered of 500 Hz)
3. To start turning and monitoring the in-process cutting forces which consist of the feed force, the main force and the radial force.
4. To examine the relation between the surface roughness and the cutting forces in the time and frequency domains.
5. To calculate the cutting forces with the personal computer.
6. To analyze the effect of cutting conditions on the surface roughness by plotting the relations of them.

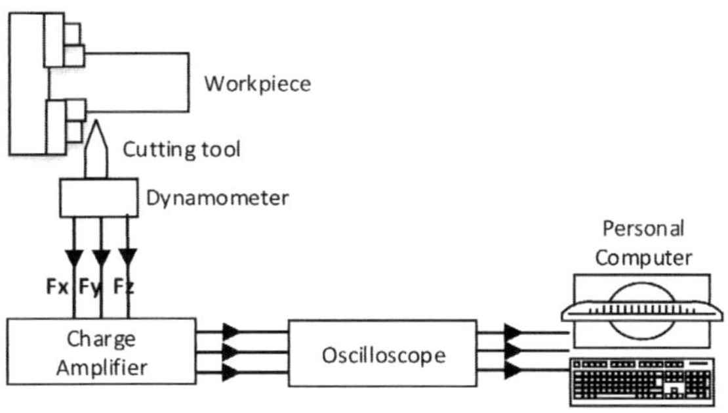

Figure 1. Illustration of experimental set up.

Table1. Cutting conditions

Cutting tool	Tool geometry	Tool nose radius (mm)	Workpiece	Cutting speed (m/min)	Feed rate (mm/rev)	Depth of cut (mm)
Coated carbide tool	TPMR160304 HQ TPMR160308 HQ	0.4 0.8	Aluminium (Al 6063)	150 200 250	0.100 0.125 0.150	0.1 0.2 0.3

4. Experimental Results and Discussions

4.1 Relation of Surface Roughness Profile and Cutting Forces

The relation between surface roughness profile and dynamic cutting forces which corresponds with time record of the surface roughness profile in time domain as shown in figure 2. It is explained that the cutting force signals are resembling the surface roughness profile. Figure 3 shows the frequencies of the dynamic cutting forces and the surface roughness profile which are the same frequency at 16 Hz by applying the Fast Fourier Transform (FFT). It is also proved that the dynamic cutting forces conform to the surface roughness profile. Thus, the dynamic cutting forces are able to predict the surface roughness during the turning.

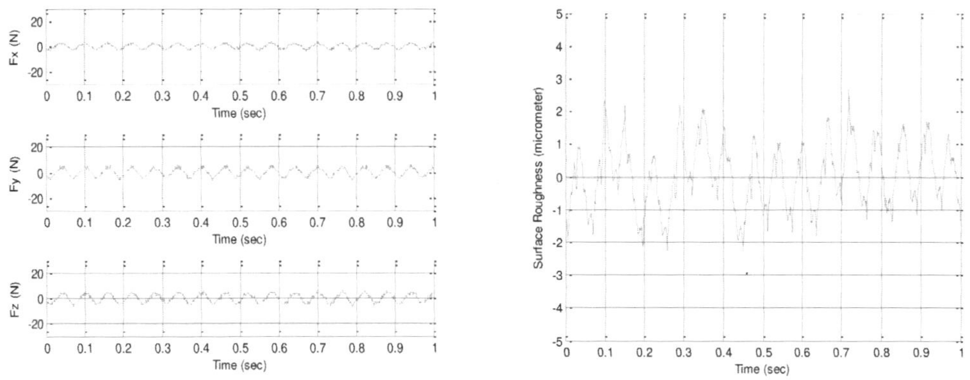

(a) Cutting forces (b) Surface roughness

Figure 2. Illustration of experimentally obtained dynamic cutting forces and surface roughness profile in time domain at cutting speed of 150 m/min, feed rate of 0.125 mm/rev, depth of cut of 0.1 mm and tool nose radius of 0.8 mm.

IOP Conf. Series: Materials Science and Engineering **303** (2017) 012013 doi:10.1088/1757-899X/303/1/012013

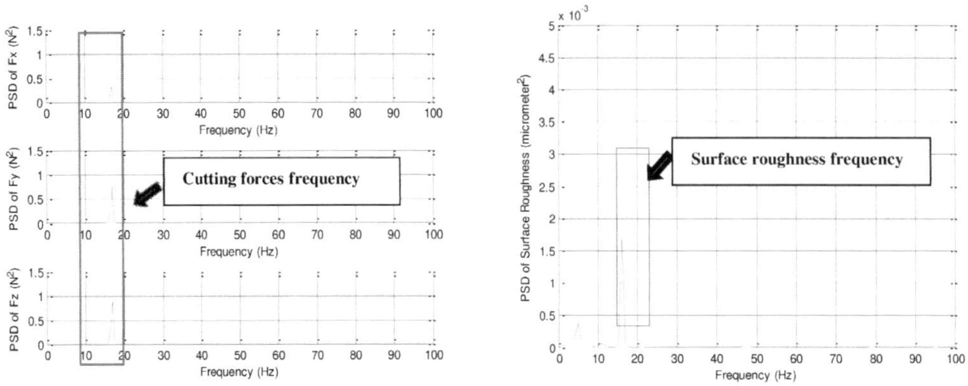

(a) Cutting forces (b) Surface roughness profile

Figure 3. Illustration of experimentally obtained dynamic cutting forces and surface roughness profile in frequency domain at cutting speed of 150 m/min, feed rate of 0.125 mm/rev, depth of cut of 0.1 mm and tool nose radius of 0.8 mm.

4.2 Effects of Cutting Conditions on Surface Roughness

Figure 4 shows the relation between the surface roughness and the cutting speed in which both of them are contrary. The cutting speed increases while the surface roughness decreases because the higher cutting speed causes the higher cutting temperature. Hence, the work material becomes softer that leads to the lower cuttings forces *[7]*. The surface finish would be improved at the higher cutting speed. However, too high the cutting speed in aluminium turning causes the built-up edge *(*BUE*)* that results in the poor surface finish.

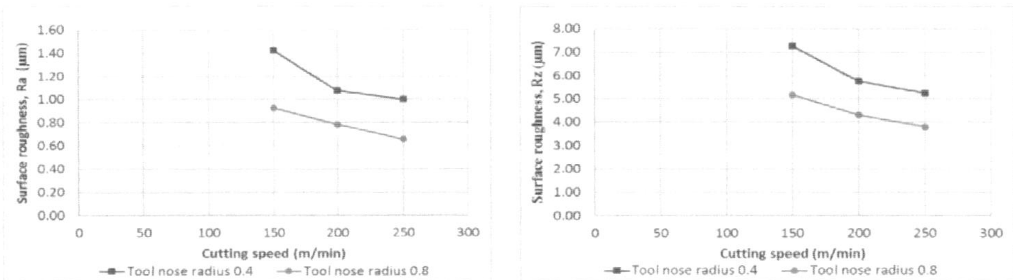

(a) Arithmetic mean surface roughness (Ra) (b) Surface roughness depth (Rz)

Figure 4. Illustration of experimentally obtained relation between surface roughness and cutting speed at feed rate of 0.100 mm/rev, depth of cut of 0.3 mm, tool nose radius of 0.4 and 0.8 mm.

Figure 5 shows that the different three feed rates comparing with the same tool nose radius, the depth of cut and the cutting speed. In this case, an increase in feed rate brings the poor surface roughness which corresponds to the theoretical surface finish. The higher feed rate causes the larger cutting area that leads to the higher cutting forces. The cutting tool will be vibrated which can result in pronounced waviness on the workpiece surface. Consequently, the surface finish is poor [9].

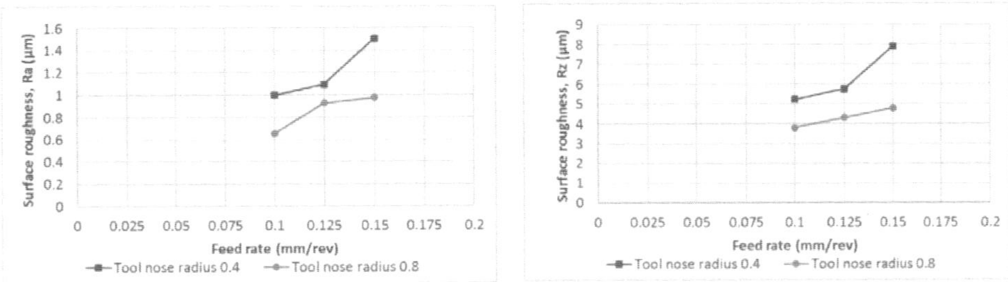

(a) Arithmetic mean surface roughness (Ra) (b) Surface roughness depth

Figure 5. Illustration of experimentally obtained relation between surface roughness and feed rate at cutting speed of 250 mm/min, depth of cut of 0.3 mm, tool nose radius of 0.4 and 0.8 mm.

Figure 6 represents the relation between the surface roughness and the depth of cut which are consistant. An increase in depth of cut will go up to the surface roughness. These are a raised cutting area which means that giving the larger of cutting forces resulting in an increase the surface roughness. However, the vibration of cutting tool during cutting is involved by the high depth of cut.

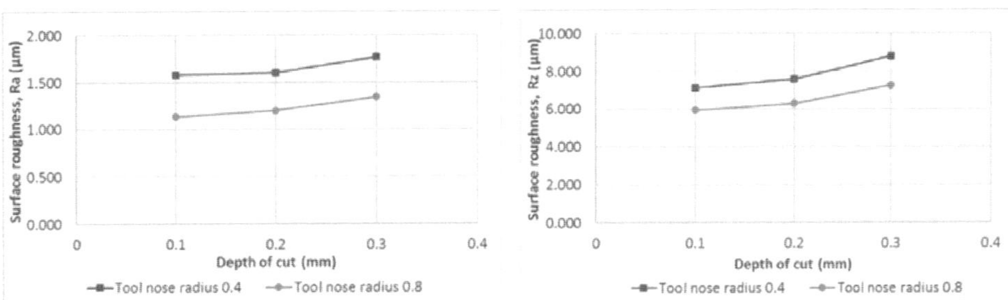

(a) Arithmetic mean surface roughness (Ra) (b) Surface roughness depth (Rz)

Figure 6. Illustration of experimentally obtained relation between surface roughness and depth of cut at cutting speed of 150 mm/min, feed rate of 0.15 mm, tool nose radius of 0.4 and 0.8 mm.

From the principles of surface roughness, it is widely known that the tool nose radius affects the quality of surface finish. With the same feed rate, the better surface finish of the tool nose radius of 0.8 mm can be obtained as compared to the tool nose radius of 0.4 mm as shown in figure 7. It can be concluded that the larger tool nose radius helps to improve the surface finish because it can reduce the feed marks on the turned surface.

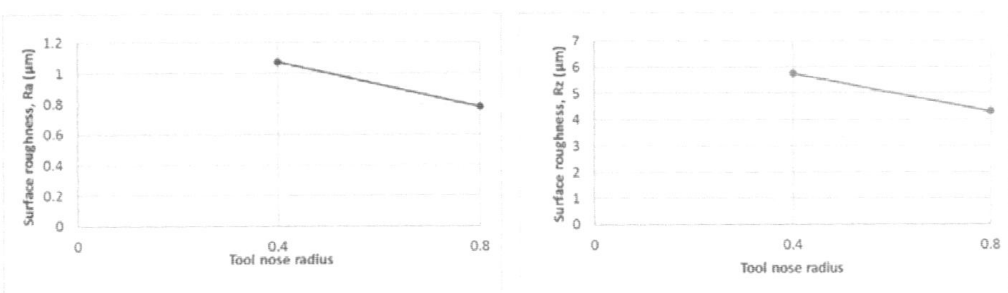

(a) Arithmetic mean surface roughness (Ra) (b) Surface roughness depth (Rz)

Figure 7. Illustration of experimentally obtained relation between surface roughness and tool nose radius at cutting speed of 200 mm/min, feed rate of 0.1 mm and depth of cut of 0.3 mm.

5. Conclusions

The relation among the surface roughness, the cutting condition and the cutting forces is studied. Based on the experimental results, the dynamic cutting forces are correspondent with the in-process surface roughness in time and frequency domains. The cutting results have demonstrated that the surface roughness become lower when the feed rate and the depth of cut are low which is in contrast to the cutting speed and the tool nose radius. Since the cutting forces are decreased by a decrease in the feed rate and the depth of cut, the work material is soften by an increase in the cutting speed. The large tool nose radius can reduce the feed marks on the turned surface that it helps to improve the surface finish.

It is implied that the in-process cutting forces are able to predict the surface roughness during the aluminium turning process.

References

[1] Thamizhmanii, S., S. Saparudin, and S. Hasan (2007) Analyses of surface roughness by turning process using Taguchi method. *Journal of Achievements in Materials and* Manufacturing *Engineering*, **20**, 503-506.

[2] Rao, C.J., D.N. Rao, and P. Srihari (2013) Influence of Cutting Parameters on Cutting Force and Surface Finish in Turning Operation. *Procedia Engineering*, **64**, 1405-1415.

[3] Lalwani, D.I., N.K. Mehta, and P.K. Jain (2008) Experimental investigations of cutting parameters influence on cutting forces and surface roughness in finish hard turning of MDN250 steel. *Journal of Materials Processing Technology,* **206**, 167-179.

[4] Thammasing, V. and S. Tangjitsitcharoen (2014) In-Process Prediction of Surface Roughness in Grinding Process by Monitoring of Cutting Force Ratio. *Applied Mechanics & Materials,* (**627**).

[5] Tangjitsitcharoen, S., K. Samanmit, and S. Ratanakuakangwan (2014) Development of surface roughness prediction by utilizing dynamic cutting force ratio. *Applied* Mechanics *and Materials,* **490**: p. 207.

[6] Tangjitsitcharoen, S. and S. Damrongthaveesak (2013) Advance in monitoring and process control of surface roughness. *World Academy of Science, Engineering and Technology,* **7**, 07-29.

[7] Tangjitsitcharoen, S. (2010) In-process prediction of surface roughness by utilizing the cutting force ratio. *Trans NAMRI/SME*,. **38**, .307-315.

[8] Serope Kalpakjian, S.R.S. (2001) *Manufacturing engineering and technology 4th Edition*, USA: Prentice-Hall, Inc.

[9] Mikell P. Groover (2010) *Fundamentals of modern manufacturing : materials, processes, and* system. Vol. 4th Ed.ition, USA: John Wiley & Sons, Inc.

[10] Tangjitsitcharoen, S. (2013) Advanced prediction of surface roughness by monitoring of dynamic cutting forces in CNC turning process. *Applied Mechanics and Materials*, Trans Tech Publ.

[11] Somkiat, T., A. Somchart, and T. Sirichan. (2010) In-process monitoring and prediction of surface roughness on CNC turning by using response surface analysis. *Proceedings of the 36th International MATADOR Conference*. 2010. Springer.

IOP Conf. Series: Materials Science and Engineering **303** (2017) 012014 doi:10.1088/1757-899X/303/1/012014

The structure and capability of hot deformation Cu-1.2Cr-0.2Zr in aging state

J W Qiao[1], L Fu[1 a], Y Y Wang[1], Y H Ding[2] and X Ma[1 a]

[1]School of Material Science and Engineering, Tianjin University of Technology; Tianjin Engineering Center of Near-Net-Shape Forming Technology for Metallic Materials, , Tianjin 300384, China
[2]School of Electrical and Electronic Engineering, Tianjin Key Laboratory of Film Electronic and Communication Devices, Tianjin University of Technology, Tianjin 300384, China.

Catjxd@126.com; maxumax@163.com

Abstract. Cu-Cr-Zr alloy belongs to precipitation-enhanced, high-strength and high-conductivity copper alloys. The volume fraction of specimen deformation was 25% and the deformation rates were 0.01~1s-1 with the deformation temperature ranged from 500°C to 800 °C at Gleeble-1500 thermal simulator. The results state that the rheological stress of Cu-Cr-Zr alloy decreases with a rising temperature at a same deformation quantity. Along with the increasement of the strain rates and the temperatures during hot deformation, the grains of Cu-Cr-Zr alloy were elongated perpendicularly to the direction of compression. When the deformation temperature was 600℃, some of fine recrystallized grains appeared.

1. Introduction

Copper and copper alloys have the most important characteristics of high electrical conductivity and thermal conductivity. With the exception of silver, copper is the most conductive in all metals, which determines its widely application and high consumption. In addition, copper and copper alloys possess excellent corrosion resistance, anti-bacterial effect, easy processing, the preparation of various alloys, etc. Copper alloys are mainly used in electronics, electric power, chemical industry, communication, machinery, marine engineering, transportation, aerospace, building decoration and other fields [1-3].
Cu-Cr-Zr alloy is a precipitation-enhanced high-strength and high-conductivity copper alloy. Many researches in the heat treatment process have been studied before, such as room temperature mechanical properties, electrochemical properties, precipitation microstructure, hot deformation behavior, processing map, ect [4-5]. However, the study of the structure and capability of hot deformation Cu-Cr-Zr alloy was less.

2. Experimental

The composition of materials used in the experiment was Cu-1.2Cr-0.6Zr alloy rolled rod with diameter of 20 mm. The raw materials were cut into a cylindrical with φ8mm × 12mm. The specimens were heated 450℃ in vacuum atmosphere, kept for 4 hours, and air cooling for aging. Hot deformation was carried out at the Gleeble-1500 thermo-mechanical simulator with parameters of table 1. The samples were polished and etched with a solution of $FeCl_3$(5g)+HCl(10ml)+H_2O (100ml).

The phase composition were testing on Rigaku D/max/2500PC X ray diffraction. And the microstructures were observed using XJP100 metallographic microscope and QUANTA FEG 250 field emission scanning electron microscope. Moreover, energy dispersion spectrum were proved the phase and component of materials discussed here. In addition, micro-hardness indicated the ability of metal materials to resist local deformation, especially plastic deformation. It is also the basis of material hardness, which represents the deformation resistance of materials.

3. Results and discussion

3.1 Thermal Deformation Analysis of Cu-Cr-Zr Alloy

Harding and strengthening will be occurred by the recovery and recrystallization taken place in plastic deformation during the annealing. Table 1 shows the flow stress of Cu-Cr-Zr alloy at a deformation volume fraction of 25% with different deformation temperatures and strain rates. It was clear that the flow stress was decreased while the deformation temperature was increased for the growing kinetic energy. The point defect is activated and the activation energy of the cutting edge dislocation increased. The corresponding kinetic energy of the metal atom is also strengthened. The critical shear stress between the atoms is reduced, when the dislocation motion is relatively easy to occur. And the dynamic softening effect is obtained, which reduces the flow stress of Cu-Cr-Zr alloy.

Table 1. Peak stress under different temperature and strain rate with 25% deformation of the Cu-Cr-Zr alloy

	Strain rate		
Temperature	**0.01 s-1**	**0.1 s-1**	**1 s-1**
800 ° C	56.52	77.01	100.8
700 ° C	96.54	134.9	167.9
600 ° C	239.7	289.6	306.7
500 ° C	344.7	365.3	379.1

3.2 XRD analysis of Cu - Cr - Zr alloy

The only peaks of Cu, absent of Cr and Zr peaks, state low content in the samples (Figure.1). Besides, it is seen that the peaks of specimens shifted to left for Cr dissolved into the Cu matrix during solution[6]. After hot deformation of the samples, thermal compression actually, the crystallinity declined and the grain size decreased for peaks broad taken on these pictures.

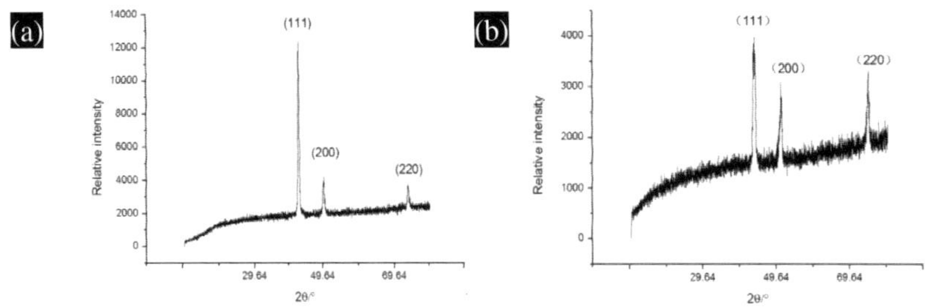

Figure1. XRD spectrum of Cu - Cr - Zr alloy with rolling state (a) and thermal deformation conditions: 800℃, deformation rate: 1s-1, deformation:25%(b)

3.3 SEM and EDS analysis of Cu - Cr - Zr alloy

In order to determine the structure, morphology, phase, composition of the alloy, SEM and EDS analysis were shown in Figure.2. The main components of the alloy were Cu, Cr, Zr.

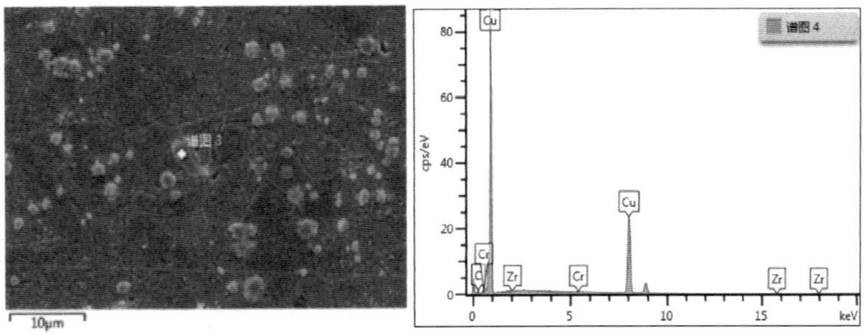

Figure 2. SEM and EDS spectrum of Cu-Cr-Zr alloy

3.4 The microstructures of Cu - Cr - Zr alloy after thermal deformation

The microstructures of Cu-Cr-Zr alloy at 800° C with different strain rates (0.01 s-1, 1 s-1, 1 s-1)were shown in Fig.3. The grain size was decreased clearly with a high strain rate for lacking enough recrystallizing time. At a large strain rate, there are not enough time for the dislocations in Cu-Cr-Zr alloy to slip while the proliferation and entanglement of dislocations were occurred, which resulting in inhomogeneous deformation, local rheological deformation and shear band. When the strain rate was $0.1s^{-1}$, the dislocations cross-slip were taken place and the grains of Cu-Cr-Zr alloy were elongated perpendicularly to the direction of compression.

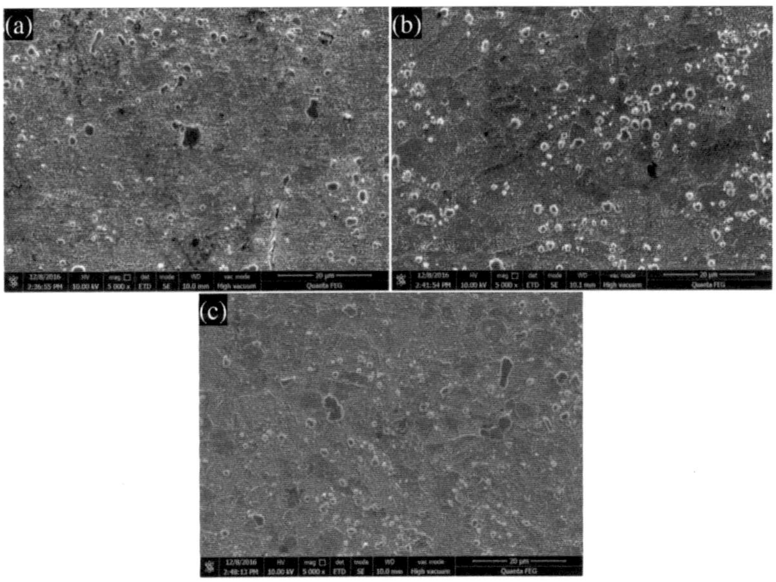

Figure 3. SEM at the thermal deformation conditions: 800 °C， deformation rate: (a) 0.01 s^{-1} (b) 0.1 s^{-1}、 (c) 1 s^{-1} when deformation: 25%

The microstructures in different thermal deformation temperature of 800°C, 700°C, 600°C, 500°C at a strain rate of $0.01s^{-1}$ were shown in Fig.4. At 500°C, obviously,the grains of the Cu-Cr-Zr alloy were compressed to fibrous structure along with the compression direction. However, recrystallization was not occured for the absence of nucleation grains. When the deformation temperature was 600 °C, some of fine recrystallized grains appeared near the original grains, which indicated that the dynamic recrystallization was generated. When the deformation temperature reached to 700℃or 800°C,the recrystallization grains enlarged obviously, which indicated that the Cu-Cr-Zr alloy underwent dynamic recrystallization. In general, there were indications that both of the grain size of the alloy and the quantity of recycled grains increase with a rising deformation temperature. Recrystallization plays a leading role in this process.

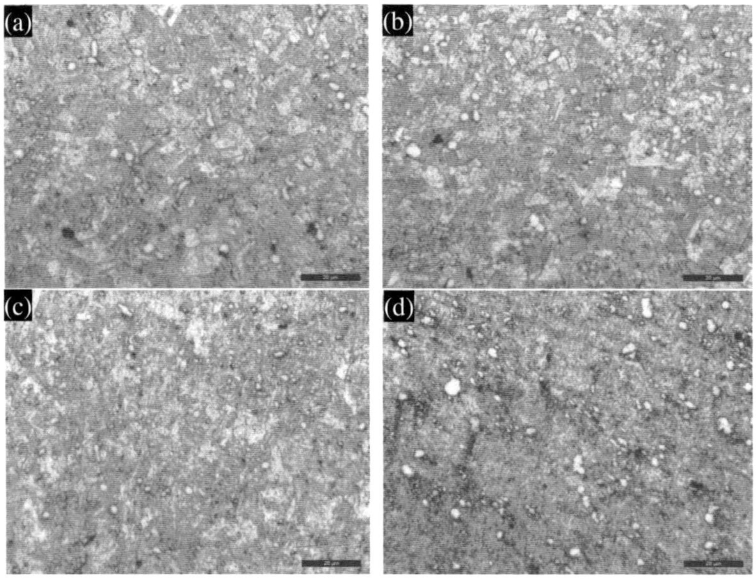

Figure 4. Metallographic at the deformation rate: 0.01of different temperature: (a) 800 °C (b) 700 °C (c) 600 °C (d) 500 °C

3.5 The micro-hardness of Cu - Cr - Zr alloy after thermal deformation

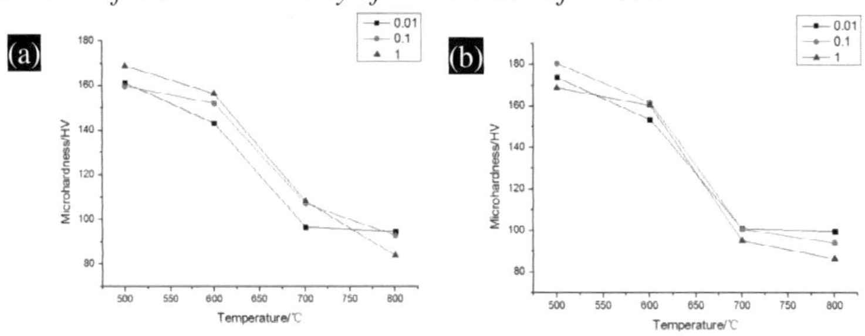

Figure.5 The change of the hardness value at the different temperature when the deformation conditions: (a)25%(b)50%

Micro-hardness refers to the resistance to local deformation especially to the plastic deformation, indentation or scratch. It is the basis of characterizing the hardness and deformation resistance of the materials. Figure.5 shows the values of hardness at different temperatures and deformation rates. It can be seen that the hardness decreases with the increase of the deformation temperature. The effect of dynamic recovery, dynamic recrystallization and work hardening of Cu-Cr-Zr alloy at different temperatures are distrinct. With gradually increasing deformation temperatures, the deformation resistance is reduced and the dynamic recrystallization occurs, which are beneficial to softening.

4. Conclusions

(1) After hot deformation of Cu-Cr-Zr alloy the crystallinity declined and the grain size decreased.
(2) The grain size was decreased clearly with a high strain rate for lacking enough recrystallizing time.
(3) Recrystallization begins at 600°C. And some of fine recrystallized grains appeared near the original grains.When the deformation temperature reached to 700℃or 800°C,the recrystallization grains enlarged obviously.
(4) With gradually increasing deformation temperatures, the values ofthe micro-hardness decreased.

References

[1] Ding Z, Jia S, Zhao P, et al. Hot deformation behavior of Cu–0.6Cr–0.03Zr alloy during compression at elevated temperatures[J]. Materials Science & Engineering A, 2013, 570(5):87-91.
[2] Zhang S.J, Li R.G, Kang H.J, et al. A high strength and high electrical conductivity Cu-Cr-Zr alloy fabricated by cryorolling and intermediate aging treatment[J]. Materials Science & Engineering A, 2017, 680: 108-114.
[3] Fu H.D, Xu S, Li W, et al. Effect of rolling and aging processes on microstructure and properties of CuCr-Zr alloy[J]. Materials Science & Engineering A, 2017, 700: 107-115.
[4] Joo H.S, Kim Y.N, Hwang S.K, Im Y.T. The effect of wire drawing and aging on mechanical and electrical properties of Cu-Cr-Zr alloy[J]. Procedia Engineering, 2017, 207: 1129-1134.
[5] Liu Y.L, Zhou P, Liu S.H, Du Y. Experimental investigation and thermodynamic description of the Cu-Cr-Zr system[J]. CALPHAD: Computer Coupling of Phase Diagrams and Thermochemistry, 2017, 59: 1-11.
[6] Zhou J.M, Zhu D.G, Tang L.T, et al. Microstructure and properties of powder metallurgy Cu-1%Cr-0.65%Zr alloy prepared by hot pressing[J]. Vacuum, 2016, 131:156-163.

Acknowledgments

This work was supported by the National Natural Science Foundation of China (No. 51101113), Tianjin Natural Science Foundation (15JCZDJC38700).

Underwater Laser Micromilling of Commercially-Pure Titanium Using Different Scan Overlaps

Wisan Charee [1*] and Viboon Tangwarodomnukun[2]

[1]Department of Industrial Engineering, Faculty of Industry and Technology, Rajamangala University of Technology Isan, Sakon Nakhon 47160, Thailand
[2]Department of Production Engineering, Faculty of Engineering, King Mongkut's University of Technology Thonburi, Bangkok 10140, Thailand

wisan.ch@rmuti.ac.th

Abstract. Underwater laser milling process is a technique for minimizing the thermal damage and gaining a higher material removal rate than processing in air. This paper presents the effect of laser scan overlap on cavity width, depth and surface roughness in the laser milling of commercially-pure titanium in water. The effects of laser pulse energy and pulse repetition rate were also examined, in which a nanosecond pulse laser emitting a 1064-nm wavelength was used in this study. The experimental results indicated that a wide and deep cavity was achievable under high laser energy and large scan overlap. According to the surface roughness, the use of high pulse repetition rate together with low laser energy can promote a smooth laser-milled surface particularly at 50% scan overlap. These findings can further suggest a suitable laser micromilling condition for titanium in roughing and finishing operations.

1. Introduction

Fabrication of micro/nano-components has become a significant attention in many fields of manufacturing industries. Lasers have been recognized as a highly potential tool to produce micro parts due to their high resolution and low processing cost compared to other technologies. However, heat generated and accumulated in work material during laser machining processes adversely introduces side effects to the workpiece in terms of heat-affected zone and recast deposition. These undesired features are usually found alongside the machined area [1, 2]. Underwater laser machining process is therefore an alternative technique to reduce such problems and also gain better cut quality as well as higher machining rate than the laser ablation in ambient air [3]. Laser milling process in water has comprehensively been studied in recent years with an attempt to understand the underlying process mechanics and its potential to fabricate the accurate and smooth micro cavities [4, 5]. However, these studies have still not provided in-depth discussion on the influence of scan overlap, especially when flowing water is applied over the workpiece surface during the ablation. Hence, this paper aims at investigating the effects of the laser scan overlap as well as other laser parameters on the cavity dimensions and surface roughness obtained in the underwater laser milling process.

2. Experimental Design and Procedure

A commercially-pure titanium (cp-Ti) sheet having the thickness of 1.0 mm was used as a workpiece sample in this study. The chemical compositions of cp-Ti are listed in Table 1. The whole workpiece was held in a stainless steel chamber as shown in Fig. 1(a). This chamber had a window on the top side allowing a laser beam to reach the workpiece surface for ablation. Inlet and outlet channels positioned on each side of the chamber introduced a directional flow of water across the top workpiece surface. The distance between the window and specimen was kept constant at 5 mm. A centrifugal pump was used to supply water into the chamber with the rate of 4 l/min.

Table 1. Chemical compositions of cp-Ti sheet

Element	Content (%)
C	0.10
Fe	0.30
H	0.015
N	0.03
O	0.25
Ti	99.20

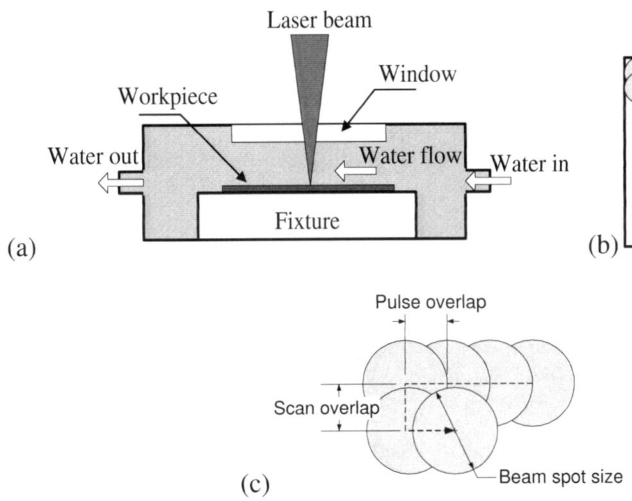

Figure 1. Schematic of (a) underwater laser milling process; (b) top view of laser milling paths; (c) pulse overlap and scan overlap

A nanosecond pulse laser providing the wavelength of 1064 nm and pulse duration of 120 ns was used in this study. The maximum laser pulse energy was 1 mJ, and the pulse repetition rate can be varied from 30 to 250 kHz. A collimated laser beam having the diameter of 10.2 mm was focused at the top workpiece surface by a 67.27-mm focusing lens. As the laser beam travelled through air, window and water before reaching the target surface, the refracted beam in water consequently caused the spot diameter of 27.11 μm at the surface [6]. This optical setup was identically employed for all tests in this study.

Process parameters considered in this research were laser scan overlap, laser pulse energy and laser pulse repetition rate. Three levels of scan overlap and two levels of laser energy and pulse repetition rate were tested in this work as shown in Table 2. A laser beam scanned over the workpiece surface with a constant traverse speed of 100 mm/min to create a 2×2 mm cavity. Each laser path overlapped the other regarding the scan overlap value, while the pulse overlap was dependent on the scan speed as illustrated in Fig. 1(c). Cavity width and depth including surface roughness were quantified from each cavity by using a high magnification confocal laser microscope (Olympus LEXT OLS4000). Surface morphology of milled cavity was qualitatively observed by a scanning electron microscope (SEM).

Table 2. Parameters used in the experiments

Process parameters	Value		
Laser scan overlap (%)	25	50	75
Laser pulse energy (mJ)	0.1	0.5	
Pulse repetition rate (kHz)	30	60	
Traverse speed (mm/min)	100		
Water flow rate (l/min)	4		

3. Results and Discussion

Fig. 2 presents the average width and depth of laser-milled cavity obtained from each test. The increase in laser pulse energy and scan overlap was found to increase the cavity width and depth. This is due to the fact that high laser energy can introduce rapid melting and vaporization of work material being irradiated, which in turn causes a large ablated area. In addition, when a laser beam scans back and forth over a workpiece surface with a large track overlap in the laser milling process, the work material located in the overlapped region undergoes reheating and more evaporation during the ablation. This consequently makes the laser-milled cavity wider and deeper under a larger scan overlap particularly with the combination of high laser energy operation. As for Fig. 2(c) and (d), the maximum depth seems to be about 175 μm. The off-focusing of laser beam could be responsible for this feature, and the light absorption in water can also attenuate the laser intensity at the workpiece surface to some extent, thus limiting the laser energy to further deepen the cut. Besides the influence of the laser energy and scan overlap, the pulse repetition rate also provided different cavity dimensions. As presented in Fig. 2, the low repetition rate can result in a slightly large and deep cut. This is on account of long pulse-off time in the low repetition rate setup that allows water to completely flush away the cut debris from the laser ablated area before firing the next laser pulse. A wider and deeper cavity can be expected in the laser ablation with low pulse repetition rate, and the similar results can be found in [4, 6, 7].

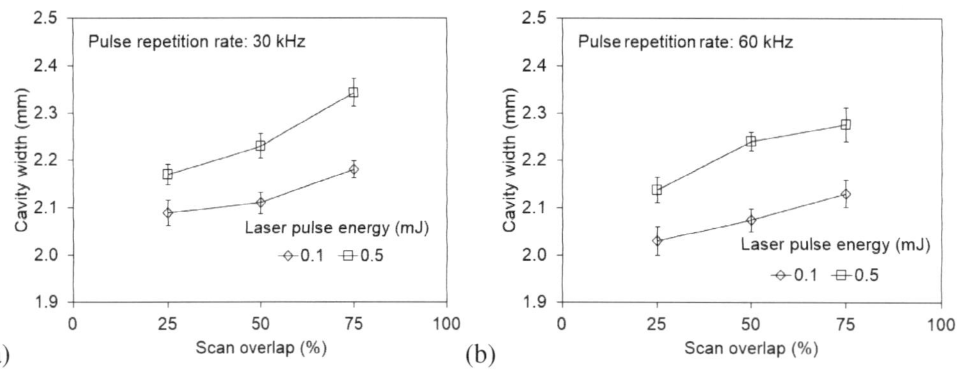

(a) (b)

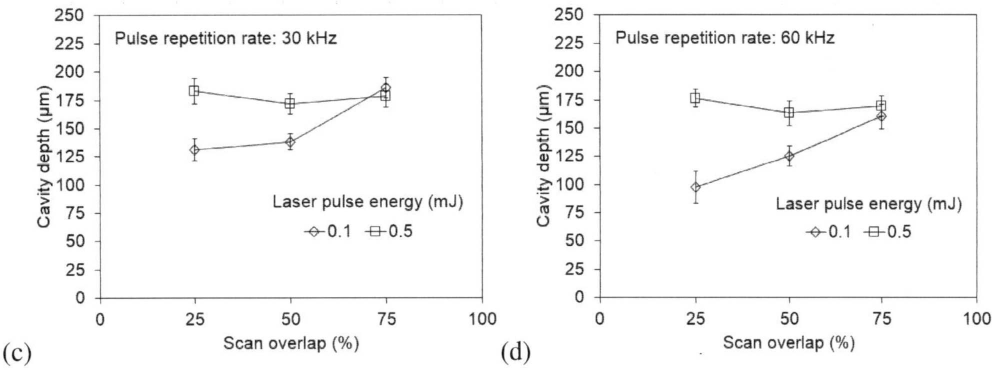

Figure 2. Effects of process parameters on cavity width and depth

Average surface roughness (Ra) of cavity was also quantified in this study as shown in Fig. 3, indicating that the use of low pulse repetition rate and high laser energy produced rough cavity surface. This is because high heat input induced by such conditions can cause large bubble formation at the laser-irradiated area in water, so that shock waves generated during the formation and collapse of bubbles can impinge toward the laser-ablated surface[8] and thereby make the surface roughened [4,9]. Since the shock waves were occurred while the laser-irradiated surface was still being the liquid or soft-solid status, the mechanical shock pressure was therefore of high potential to deform the work material or even create small holes in the cavity surface as shown in Fig. 4. With regard to the experimental results, the scan overlap of 50% was able to provide the smoothest cavity surface among the others. When too small overlap is applied in the laser milling process, the large distance between consecutive milling tracks and weak laser intensity around the rim of Gaussian laser spot cannot effectively facilitate the ablation of material elements in-between milling tracks. This thus leads to a significant peak-to-valley surface roughness so as the average roughness value. On the contrary, too large overlap causes more amount of molten material at the cavity surface whose morphology can easily be altered by the mechanical shock pressure induced by bubbles in water, thus deteriorating the laser-milled surface quality. The average surface roughness of about 0.4 µm was achievable when 50% scan overlap was applied together with 0.1-mJ laser pulse energy and 60-kHz repetition rate. This condition can be used for surface finishing in the underwater laser milling process.

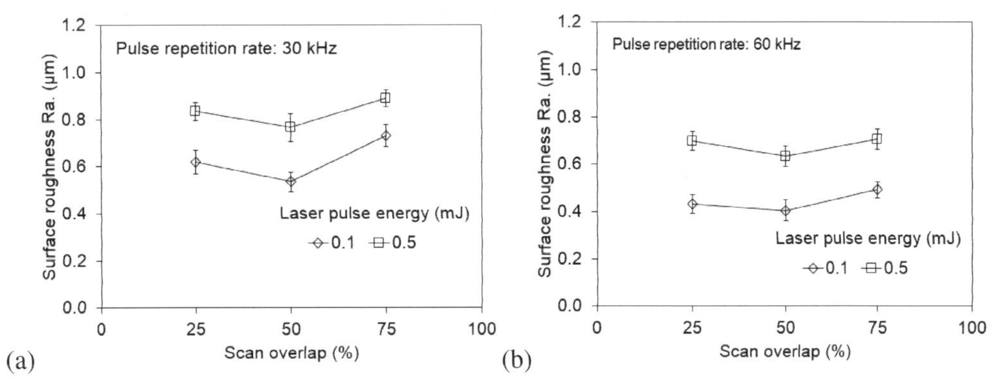

Figure 3. Effect of process parameters on surface roughness

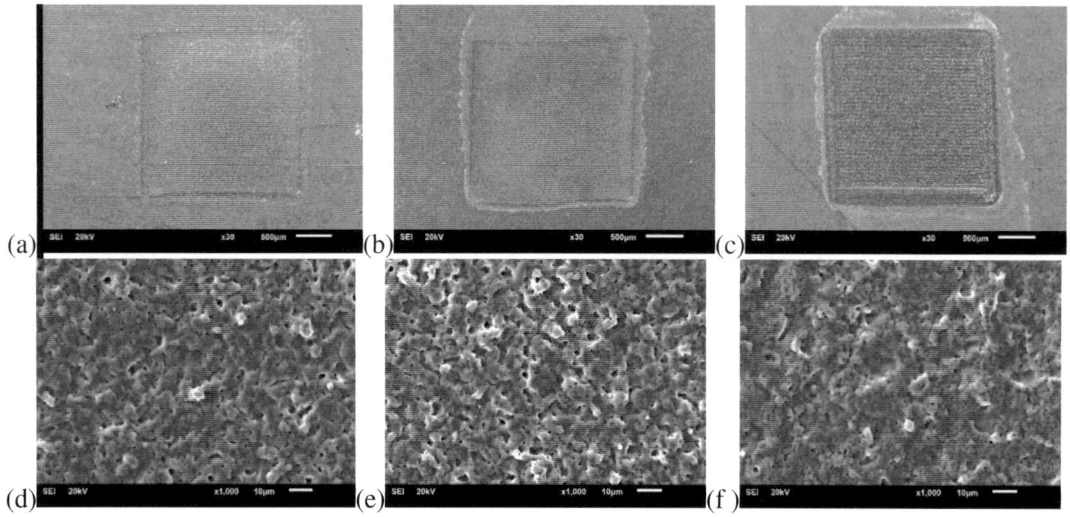

Figure 4. Surface morphology of titanium induced by the underwater laser milling process with the (a,d) scan overlaps of 25%, (b,e) 50% and (c,f) 75% when the laser pulse energy of 0.1 mJ and pulse repetition rate of 30 kHz were used

4. Conclusions

This study investigated the influence of laser scan overlap as well as laser pulse energy and laser pulse repetition rate in the underwater laser milling of commercially-pure titanium in flowing water. The width, depth and surface roughness of cavities were measured and discussed in this paper. The major findings and implications of this work can be concluded as follows:

- A wide and deep cavity was obtained when high laser pulse energy, low pulse repetition rate and large scan overlap were applied in the laser micromachining process in water. This condition could be used for rough cutting operation, which the high material removal rate is the primary concern. However, the compensation of milling paths is a must to yield the accurate cavity's dimensions.
- The smooth milled surface can be produced by using 50% scan overlap together with low laser pulse energy and high pulse repetition rate. Though this setup is likely suitable for finishing operation, the decreased ablation rate is to be sacrificed.
- Laser-milled surface improvement and micro-fabrication of three dimensional features in biomedical materials will further be investigated to unlock the potential of underwater laser micromilling process for manufacturing practice.

5. Acknowledgements

This research was financially supported by the Rajamangala University of Technology Isan. This work was also supported by the Thailand Research Fund and the Office of the Higher Education Commission, Ministry of Education, Thailand (Grant No. MRG6080010).

6. References

[1] Leone C, Papa I, Tagliaferri F and Lopresto V 2013 *Composites Part A: Applied Science and Manufacturing* **55** 129-42

[2] Marla D, Bhandarkar UV and Joshi SS 2013 *Manufacturing Letters* **2** 13–16

[3] Tangwarodomnukun V, Wang J and Mathew P 2010 *Key Engineering Materials* **443** 693-698

[4] Tangwarodomnukun V 2016 *Applied Surface Science* **386** 51–64

IOP Conf. Series: Materials Science and Engineering **303** (2017) 012015 doi:10.1088/1757-899X/303/1/012015

[5] Wuttasarn T, Tangwarodomnukun V and Dumkum C 2015 *Proceedings of the Institution of Mechanical Engineers, Part B: Journal of Engineering Manufacture* 1-5

[6] Charee W, Tangwarodomnukun V and Dumkum C 2015 *International Journal of Advanced Manufacturing Technology* **78** 19-29

[7] Charee W, Tangwarodomnukun V and Dumkum C 2016 *Applied Mechanics and Materials* **835** 144-148

[8] Charee W and Tangwarodomnukun V 2018 *Optics & Laser Technology* **100** 230-243

[9] Charee W and Tangwarodomnukun V 2016 *Materials Science Forum* **861** 115-120

IOP Conf. Series: Materials Science and Engineering **303** (2017) 012016 doi:10.1088/1757-899X/303/1/012016

Forecasting of global solar radiation using anfis and armax techniques

Auwal Muhammad[1, a], M. S. Gaya[2,b*], Rakiya Aliyu[1], Rabi'u Aliyu Abdulkadir[2], Ibrahim Dauda Umar[1], Lukuman Aminu Yusuf[3] and Mudassir Umar Ali[1],M.T.M Khairi[4]

[1]Department of Physics, Kano University of Science and Technology, Wudil, Nigeria
[2]Department of Electrical Engineering, Kano University of Science and Technology, Wudil, Nigeria
[3]Department of Electrical Engineering, Bayero University, Kano, Nigeria
[4]Deparment of Control & Mechatronics, Universiti Teknologi Malaysia, Malaysia

[a]auwalsafa@gmail.com, [b]muhdgayasani@gmail.com

Abstract. Procurement of measuring device, maintenance cost coupled with calibration of the instrument contributed to the difficulty in forecasting of global solar radiation in underdeveloped countries. Most of the available regressional and mathematical models do not capture well the behavior of the global solar radiation. This paper presents the comparison of Adaptive Neuro Fuzzy Inference System (ANFIS) and Autoregressive Moving Average with eXogenous term (ARMAX) in forecasting global solar radiation. Full-Scale (experimental) data of Nigerian metrological agency, Sultan Abubakar III international airport Sokoto was used to validate the models. The simulation results demonstrated that the ANFIS model having achieved MAPE of 5.34% outperformed the ARMAX model. The ANFIS could be a valuable tool for forecasting the global solar radiation.

1. Introduction

Global solar radiation refers to the rate of the algebraic sum of direct and diffuses incoming solar energy received at the earth's surface. Pyranometer is the most commonly used instrument to measure global solar radiation. Procurement of the pyranometer instrument, its maintenance cost together with calibration of the instrument made the forecasting of global solar radiation a quite difficult and challenging task.

Literature studies indicated that several models ranging from intelligent models to regressional models were developed. Regressional models such as ARMA and ARX belong to a family of linear models, commonly used in forecasting due to the capability of dealing with large data samples and may yield a better forecasting, but produce poor forecasting for few samples of data [1]. Intelligent models such as neural network [2], [3], [4] and Fuzzy logic are versatile, accurate and effective in handling noisy/ few samples data. However, choice of structure, trapping in local minima and selecting of membership function are the limitations of intelligent models.

A combination of neural network and fuzzy logic yielded a neuro-fuzzy which overcomes the limitations of the individual method. ANFIS belongs to a class of hybrid neuro-fuzzy and has received universal acceptability since evolution. ANFIS has an effective capability for nonlinear mapping. Linear models such as ARMAX are straightforward to implement and frequently used as a traditional tool for time-series forecasting [1]. ARMAX is an extension of ARX model and more versatile in handling disturbance modelling compared to ARX model. This paper centered on comparing ANFIS and ARMAX in forecasting the global solar radiation.

The forecasting performances of the models were evaluated using mean absolute percentage error (MAPE) and root mean square error (RMSE). Based on the judgement as reported in [5], model with MAPE less than or equal to 10% is considered highly accurate. The paper is organized as section 2 describes the study area, section 3 discusses ANFIS and ARMAX models while section 4 presents the simulation results and discussion.

2. Study Area

Nigeria is located in the equatorial region between latitudes 4oN and 14oN, longitudes 3oE and 15oE, the annual average daily sunshine ranges from about 3.5 hours at the coastal areas to about 9.0 hours at the far northern boundary, while the annual average daily solar radiation for the location is between 3.5 kWhm-2day-1 and 7.0 kWhm-2day-1 respectively [6]. Sokoto state is found in the North Western region of Nigeria, and is located in the sudano sahelian savanna ecological belt of Nigeria with longitude 11o3' to 13o50'E, latitude 4o to 6o40'. Sokoto falls within the hottest region in Nigeria. Therefore, the temperatures are of high extremes, with an average daily minimum of 16oC during cool months of January and December, and similarly in the hottest months of April to June, an average maximum of 38oC and minimum of 24oC [7, 8].

In this paper, meteorological data consisting of the monthly mean of global solar radiation, sunshine duration, maximum temperature, wind speed, rainfall, relative humidity and mean evaporation for a period of 10 years (2005-2014) for Sokoto, Northwestern Nigeria, with latitude of 12.55°N and longitude of 5.0°E were used. The data obtained from Nigerian Meteorological Agency, Sultan Abubakar III International Airport Sokoto, Sokoto State. The four parameters serve as input parameters for predicting solar radiation.

3. Methodology

ANFIS: ANFIS has an ability of fast learning, adaptability, effective handling of imprecision and uncertainty [9]. ANFIS is composed of nodes and five layers as illustrated in Fig. 1 below [10]. The square nodes are known as adaptive having parameters in them to be updated during learning process, and the circular nodes are fixed.

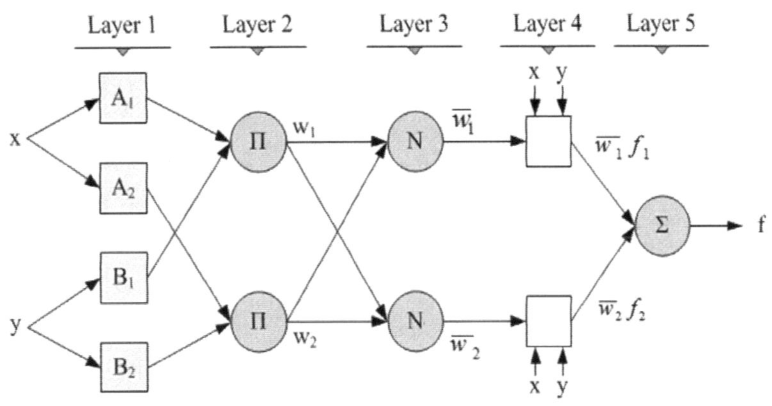

Figure 1. ANFIS structure

The structure of ANFIS is not unique, two or more layers could be combined together and produce same results [9]. The parameters of ANFIS are updated through supervise learning to reduce the error measure.

ANFIS Model: Since historical data may contain corrupted measurements, missing values, erroneous values and outliers. The data were treated in order to achieve reliable prediction. The data were selected randomly into training and testing data, 90% for training and 10% for testing. Two (2) Gaussian membership functions were assigned to each input, which generated sixty four (64) fuzzy rule. The structure of ANFIS is developed using MatLab command "genfis 1" to generate first- order Sugeno Fuzzy Inference System (FIS). As the structure is realized, the ANFIS uses hybrid learning algorithm to update the parameters of the FIS through learning from the data set to realize the target.

ARMAX: ARMAX is essentially a linear regression model that utilizes an auto regressive moving average type model for residuals. The following equation depicts the form of the ARMAX model:

$$A(q)\,y(t)=B(q)u(t-nk)+C(q)e(t) \qquad (1)$$

Where $y(t)$ is the system is outputs; $u(t)$ is the system inputs, **Error! Bookmark not defined.** is the system delay and $e(t)$ is the system disturbance. $A(q)$, $B(q)$ and $C(q)$ are polynomial with respect to the backward shift operator q.

ARMAX Model. The same treated data used for developing ANFIS model were utilized for realizing ARMAX model and with 90% of the data for training and 10% for testing. The structure of the model was selected with na, nb, nc and nk equal 2 [2 2 2 2 2 2] 2 [1 1 1 1 1 1] and the prediction error method was utilized for the forecasting.

4. Results and Discussions

The Fig. 2 and Fig 3. demonstrated the inputs which include relative humidity (RH), rainfall (RF), maximum temperature (Max T), wind speed (WS), sun shine hours (SSH) and mean evaporation (MEV).

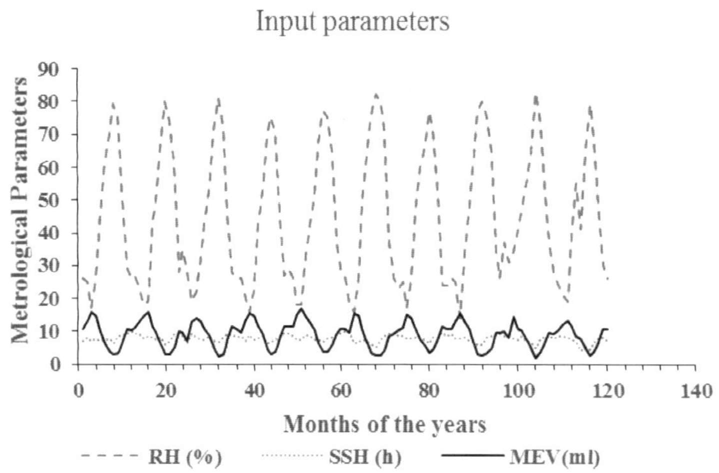

Figure 2. Input metrological parameters

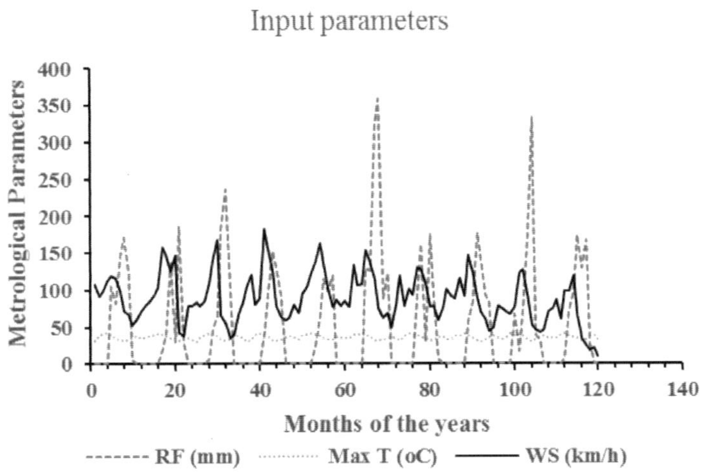

Figure 3. Input metrological parameters

The commonly available model evaluation criteria such as root mean square error (RMSE) and mean absolute percentage error (MAPE) were used measure the forecasting performances of the models.

$$MAPE = \frac{100\%}{N} * \sum_{i=1}^{N} \left| \frac{x_i - y_i}{x_i} \right| \quad (2)$$

$$RMSE = \sqrt{\frac{\sum (x_i - y_i)^2}{N}} \quad (3)$$

Where x_i depicts measured value, y_i is the forecast value, and N is the number of samples.

The Fig. 4 and Fig. 5 illustrate the forecasting performances of the model during the training and testing respectively. The Table 1 presents the results obtained during the training and testing.

Table1: Model Performance

Model	Training		Testing	
	MAPE (%)	RMSE	MAPE (%)	RMSE
ANFIS	0.002	0.001	5.340	0.562
ARMAX	45.981	2.717	41.063	2.586

It is apparent from the Fig. 4 during the training phase that the ANFIS model outperformed the ARMAX in forecasting having tracked well the pattern of the measured global solar radiation and achieved the MAPE of 0.002%, thus indicating a highly accurate forecasting.

Similarly, during testing as shown in Fig. 5, the ANFIS model predicted well the measured global solar radiation having achieved the MAPE of 5.340%, which is also highly accurate forecasting. The performances of ARMAX in both training and testing are below expectation as compared to that of ANFIS model.

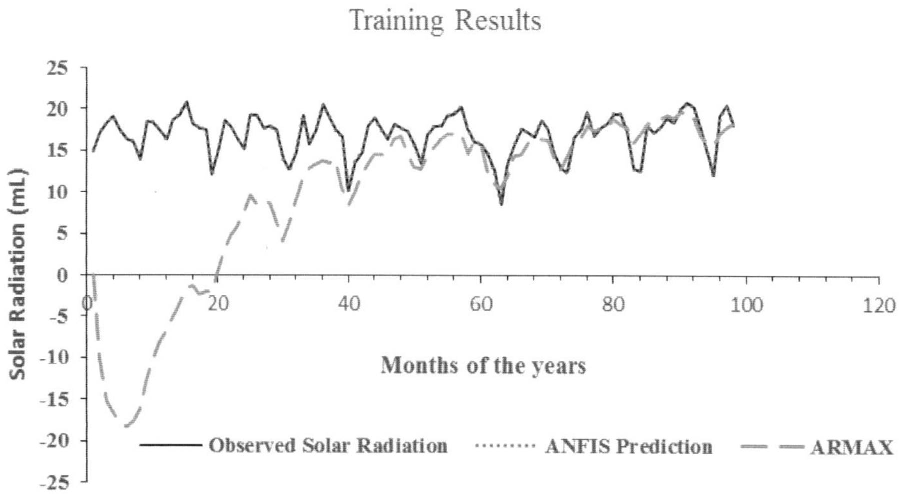

Figure 4. Models forecasting performance during training

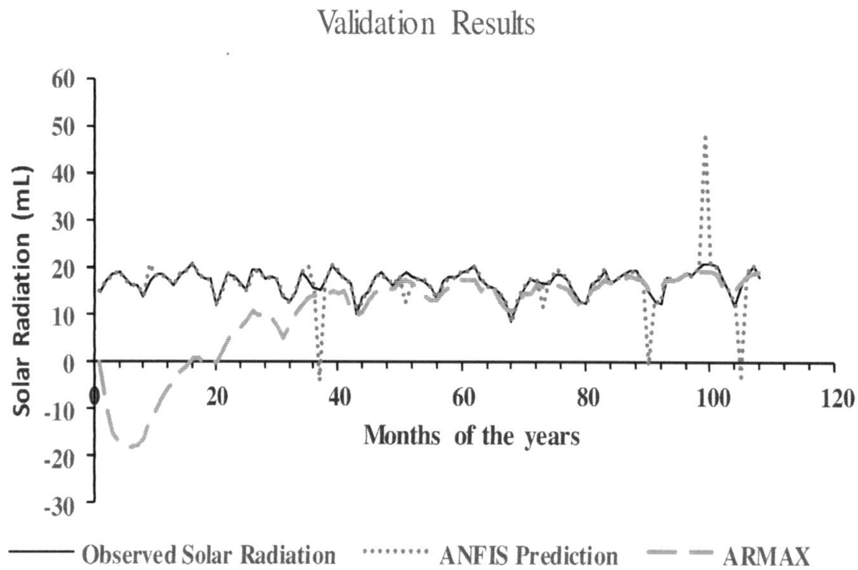

Figure 5. Models forecasting performance during testing

5. Conclusion

The paper has presented an ANFIS and ARMAX techniques in forecasting global solar radiation. The results obtained demonstrated the ANFIS performed far better than the ARMAX model having achieved the MAPE of less than 10% in both training and testing. Increasing the order of the ARMAX may enhance the performance of the model. However computational burden needs to be avoided. The ANFIS model would serve as a useful forecasting tool for the global solar radiation in the study area and those with similar geographical information.

IOP Conf. Series: Materials Science and Engineering **303** (2017) 012016 doi:10.1088/1757-899X/303/1/012016

References

[1] M. S. Gaya, N. Abdulwahab, Y. M. Sam, S. I. Samsudin, and I. W. Jamaludin: Applied mechanics and Materials Vol. 554 (2014), p. 360-365

[2] M. Benghanem, A. Mellit, and S. N. Alamri: Energy conversion and Management Vol. 50 (2009), p. 1644-1655

[3] A. Hasni, A. Sehli, B. Braoui, A. Bassou, and B. Mieur: Energy Procedia Vol.18 (2012), p. 531-537

[4] J. Mubiru and E. J. K. B. Banda: Solar energy Vol. 82 (2008), p. 181-187

[5] K. D. Lawrence, R. K. Klimberg and S. M Lawrence: *Fundamental of Forecasting using Excel* (Industrial Press Inc., New York 2009)

[6] S. Salisu, B. O. Sadiq, A. S. Abubakar, Y. Jibril and S. Garba: International Journal of Scientific and Engineering Research Vol. 5 (2014), p. 669-675

[7] S. A. Abdullahi, M. M. Mohammed, B. K. Adeoyun, and I. U. Mohammed: Resources and Enviroment Vol. 4 (2014), p. 220-233

[8] G. Atedhor: African Journal of Food Agriculture, Nutrition and Development Vol. 15 (2015), p. 9856-9871

[9] M.S Gaya: *Neuro-Fuzzy Modelling and Neural Network Internal Control of an Activated* Sludge *System* (Universiti Teknologi Malaysia PhD Thesis 2014)

[10] J. S. R. Jang: IEEE transactions on system, man and cybernetics Vol. 23 (1993), p. 665-668

Enhanced pid vs model predictive control applied to bldc motor

M.S Gaya*[1, a], Auwal Muhammad[2,b], Rabiu Aliyu Abdulkadir[1],S.N.S Salim[3], I.S Madugu[1], Aminu Tijjani[1], Lukman Aminu Yusuf[4], Ibrahim Dauda Umar[2], M.T.M Khairi[5,c]

[1]Dept of Electrical Engineering, Kano University of Science & Technology,Wudil,Nigeria
[2]Dept of Physics, Kano University of Science & Technology, Wudil
[3]Dept of Industrial Automation & Robotic, Universiti Teknikal Malaysia, Melaka, Malaysia
[4]Dept of Electrical Engineering, Bayero University Kano, Nigeria
[5]Dept of Control & Mechatronics, Universiti Teknologi Malaysia, Skudai, Nigeria

[a]muhdgayasani@gmail.com, [b]auwalsafa@gmail.com, [c]rabiukk@gmail.com

Abstract. BrushLess Direct Current (BLDC) motor is a multivariable and highly complex nonlinear system. Variation of internal parameter values with environment or reference signal increases the difficulty in controlling the BLDC effectively. Advanced control strategies (like model predictive control) often have to be integrated to satisfy the control desires. Enhancing or proper tuning of a conventional algorithm results in achieving the desired performance. This paper presents a performance comparison of Enhanced PID and Model Predictive Control (MPC) applied to brushless direct current motor. The simulation results demonstrated that the PSO-PID is slightly better than the PID and MPC in tracking the trajectory of the reference signal. The proposed scheme could be useful algorithms for the system.

1. Introduction

Recently, BLDC has received much attention due to high efficiency, long operating range and noiseless operation as compared to brushed and induction motor. Complex nonlinear systems such as BLDC whose internal parameter values vary with environment or reference signal are mostly difficult to control. Control techniques are implemented to provide the desired stability and improve performance, which may reduce energy consumption and cost. Conventional controllers like PID are effective for linear systems, but their performances deteriorate when handling systems such as BLDC.

This enticed the implementation of the different kind of control techniques to the system such as hybrid neuro fuzzy [1], adaptive fuzzy logic [2], robust H∞ fuzzy [3], robust nonlinear control [4], neural network control [5], so that the desired performance and stability can be achieved. Nevertheless, the main issues with these control techniques are the requirement of prior knowledge, high expertise and tuning complexities. Enhanced PID control is quite simple, capable of dealing with nonlinearities and environmental variation.

Several methods are used to enhance performance of PID but this research work utilized particle swarm optimization (PSO) method. PSO is straightforward and has fewer parameters to tune. Model predictive control (MPC) uses a dynamic model of a system to predict the future direction of the desired response of the system. The primary concept of MPC is the frequent selection of a control effort to solve on line an optimal control problem, and this results in minimizing the performance cost function. MPC methodology has been in used in many areas of applications.

This paper focused on comparing the performances of enhanced PID and MPC in terms set-point tracking. The performances of the controllers were evaluated based on the commonly used control measures such as rise time, settling time and percent overshoot.

2. Methodology

2.1 Plant Dynamic Model

Mostly the representation of a system either mechanistic or empirical is a key in building reliable controller. Model describes the dynamics of a system which aid significantly in developing an effective control algorithm for the system.

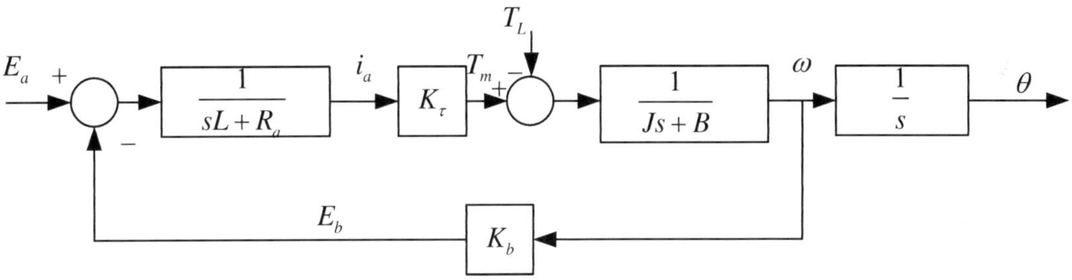

Figure 1. BLDC block diagram

From the block diagram shown in Fig. 1 using Laplace transformation the motor torque can be expressed as

$$T_m(s) = K_\tau i_a(s) \quad (1)$$

The back e.m.f is given by

$$E_b(s) = K_b \omega(s) \quad (2)$$

The mechanical load is expressed as

$$\omega(s) = \frac{1}{Js+B}\left(T_m(s) - T_L(s)\right) \quad (3)$$

The angular position is mathematically represented as

$$\theta(s) = \frac{\omega(s)}{s} \quad (4)$$

Setting $T_L = 0$, the transfer function is given by

$$\frac{\theta(s)}{E_a(s)} = \frac{K_\tau}{s\left[(sL+R_a)(Js+B) + K_\tau K_b\right]} \quad (5)$$

Where E_a is the armature e.m.f, L is the inductance, R_a is the armature resistance, J is rotor inertia, B is damping coefficient and K_τ is the torque constant. It is assumed that $L \ll R_a$ and L is set to zero in most cases.

Once the mathematical model of the plant is realized, the control algorithm can be design for the system.

2.2 Model Predictive Control

Model predictive control is a class of control algorithm that has been applied experimentally and through simulation to control several systems ranging from simple to complex, linear to nonlinear and multivariable plant. The algorithm uses the process model to predict the future output response of a system. At each control interval, the algorithm attempts to optimize future plant behaviour by computing a sequence of future manipulated variable adjustments. The first input in the optimal sequence is then sent to the plant, and the entire calculation is repeated at subsequent control intervals [6].

The predictive capability coupled with the classical feedback operation allows the algorithm to make adjustments that are smoother and closer to the optimal control action values. The controller comprises of an optimization problem at each time interval, k and the main goals are to calculate a new control signal u_k to be injected into the plant and concurrently consider the process constraints. The Fig. 2 shows the architecture of the model predictive control.

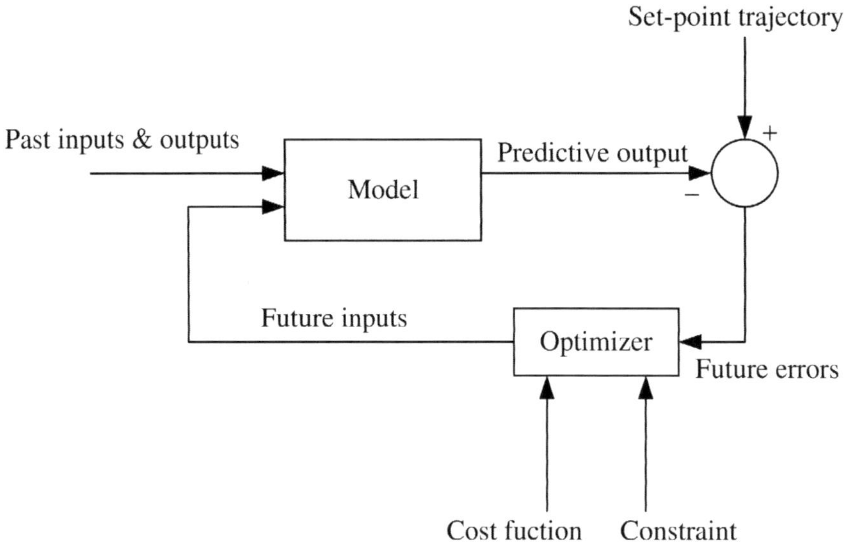

Figure 2. MPC architecture

The above Fig. 2 it is evident that the MPC is composed of cost function, constraints and model of the plant. The goal is to minimize the cost function given by

$$\Im = \sum_{k=0}^{N_P} \left(\hat{y} - r\right)^T Q \left(\hat{y} - r\right) + \sum_{k=0}^{N_P} \left(\Delta u^T R \Delta u\right) \tag{6}$$

where N_P depicts prediction horizon, r is the set-point, $\hat{y}$ is the predicted output, Δu is the predicted change in the control signal, Q and R are output error weight matrix and control weight matrix respectively.

In developing MPC algorithm, the proper choice of design parameters such as control horizon (N_C), prediction horizon (N_P), model horizon (N_M), sampling interval (Δt),output error weight matrix (Q) and control weight matrix (R) is of paramount importance as this would lead to an effective and robust controller [7].

Control horizon (N_C): This refers to the amount of manipulated variable moves to be optimized at control interval *k*. Control horizon is selected such that N_C □ N_P . Small N_C results in an internally stable controller where high (or increasing) value results in computational burden and makes the MPC more aggressive.

Prediction horizon(N_P): This refers to the amount of future control intervals the MPC controller must assess by prediction when optimizing its manipulated variables at control interval *k.* Decreasing the value of N_P makes the controller more aggressive whereas results in less aggressive. The practice is to increase N_P until further increases has a slight effect on performance.

Sampling Period: Selecting the value of sampling period is a compromise between performance and computational effort. Small value of sampling period results increase in computational burden.

Q and R matrices: Since the goal is to minimize the cost function these matrices are selected as $Q \geq 0$ (positive semi-definite symmetry matrix) and $R > 0$ (positive definite symmetry matrix).

2.3 MPC Design

Since there is no systematic technique to choose the parameters, the aforementioned criteria was utilized as the guide, after some trials and error. The controller parameters are selected as: control horizon $(N_C) = 2$, prediction horizon $(N_P) = 30$ and sampling interval $(S_T) = 0.0002$.

2.4 Enhanced PID

Several techniques to enhance PID exist, but this paper utilizes particle swarm optimization method. Particle swarm optimization is a population-based search procedure introduced by [8] in 1995. It is straightforward to implement because of few parameters to update. Appropriate selection of these parameters values are essential to achieve the desired results and speedy convergence [9]. For detailed explanation on the parameters selection could be found in [9], [10]. The flow chart in Fig. 3 illustrates the procedure for searching of the optimal gains values of PID. The PSO algorithm used are number of particles $N = 20$, Dimension $D = 3$, number of iteration $it = 50$, cognitive constant $c = 1.98$, social constant $s = 0.89$, inertia weight $iw = 0.85$.

The optimal gains of the PID controller were obtained through using the PSO algorithm shown in Fig. 3 and running the simulation on MATLAB/SIMULINK environment.

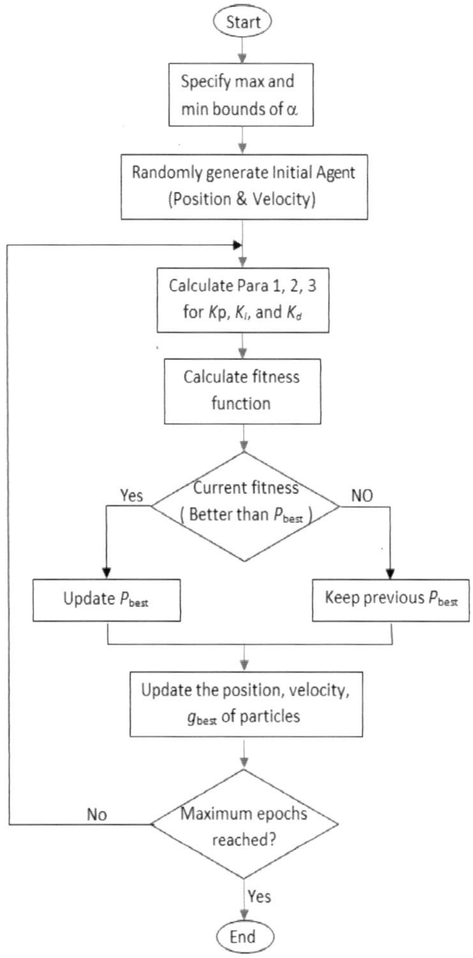

Figure 3. PSO-PID tuning flow chart

The realized PSO-PID parameters gains are $k_{p(pso)} = 0.12$, $k_{i(pso)} = 0.01$ and $k_{d(pso)} = 0.02$. Since the gains were PSO-PID obtained the performance of the controller could be assessed through simulation.

3. Simulation Results and Discussion

The simulation parameters used throughout this paper are shown in the Table 1. The open-loop response from the system indicated that error exists as illustrated in Fig. 4 when a step signal of 1radian was injected into the system.

Table 1. Simulation Parameters

Parameters	$B\ (Nms)$	$J\ (Nms^2)$	K_τ	$L_a\ (H)$	$R_a\ (\Omega)$	$i = 50$
Value	0.708×10^{-4}	11×10^{-6}	36.8	2.758×10^{-6}	4.31	$i = 50$

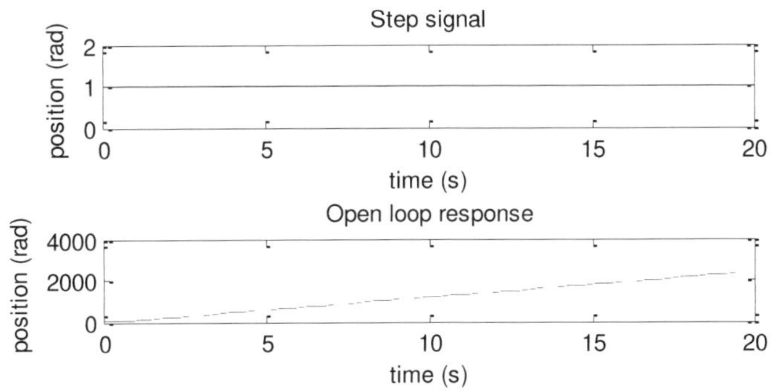

Figure 4. Open loop response

Therefore, an enhanced PID and MPC were designed as described above to minimize the error. The gains of the classical PID parameters are $k_p = 0.104$, $k_i = 0.010$ and $k_d = 0.011$. The performances of the controllers are shown in Fig.5, the commonly used control measures as mentioned earlier were utilized to evaluate the performances and the results were presented in Table 2. In addition, a variable reference signal was also used to assess the tracking capability of the controllers as depicted in Fig 6. The controllers have tracked well the reference signals, but the PSO-PID has faster response compared to the PID and MPC. Nevertheless, proper tuning of MPC may yield promising results.

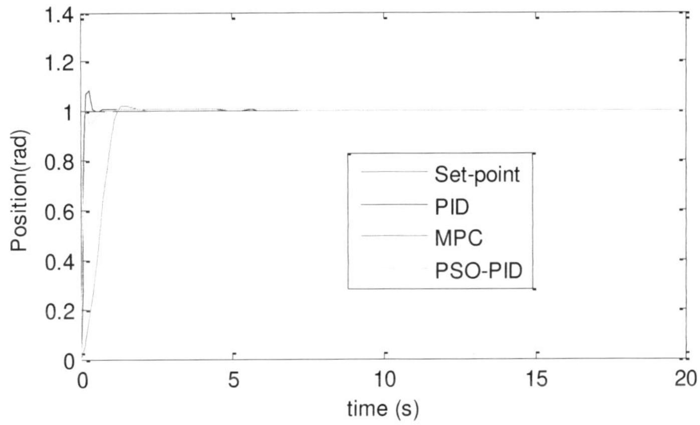

Figure 5. Controllers set-point tracking

Table 2. Controllers Performance

Control Measure	Controllers		
	PID	PSO-PID	MPC
Rise time [t_r](s)	0.1242	0.1074	0.8128
Settling time [t_s](s)	0.3972	0.5299	1.6590
Maximum overshoot [Mp](s)	8.1870	1.5297	2.3940

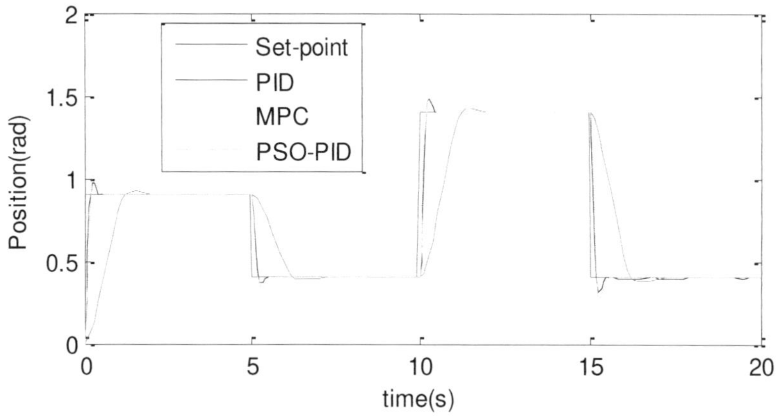

Figure 6. Variable set-point tracking performance

4. Conclusion

The paper has presented the enhanced PID and MPC for BLDC motor position control. The results obtained demonstrated that the PSO-PID performed slightly better than the classical PID and MPC. Even with the variable reference signal, the PSO-PID has better response than the MPC and PID. Although disturbance rejection capability of the controllers was not evaluated, with proper tuning of MPC, a promising result could be achieved. The proposed controllers could serve as an alternative control architecture for the brushless direct current motor.

References

[1] M. Gokbulut, B. Dandil and C. Bal: TAINN Springer Vol. 83-87 (2006), p. 125-132

[2] E. N. Binti Sabudin: *Development of Position Tracking of BLDC Motor using Adaptive Fuzzy* Logic *Controller* (Universiti Tun Hussein Onn, Malaysia, Masters Thesis, 2012).

[3] W. Assawinchaichote and N.Chayaopas: Applied Mathematics and Information Sciences Vol. 10 (2016), p. 987-995

[4] N. Hemati, J. S. Thorp and M. C. Leu: IEEE Transactions on Industrial Electronics Vol. 37 (1990), p. 460-468

[5] M. Gokbulut and A. Tekin: Int. J. Engng Ed. Vol. 22 (2006), p. 197-204

[6] H. P. Halvorsen: *Model Predictive Control in Labview* (Faculty of Technology, Porsgrunn, Norway, 2011)

[7] A. Bemporad, M. Morari and N. L. Ricker: *Model Predictive Control Toolbox for use with* MATLAB (Math Works, Inc., U. S. A., 2014)

[8] J. Kennedy and R. C. Eberhart: Proceeding of IEEE international conference on neural networks Vol. IV, pp. 1942-1948. IEEE service center, Piscataway, NJ, 1995

[9] M. S. Gaya, A. A Bisu, S. N. S. Salim, I. S. Madugu, L. A. Yusuf, M. L Kaurangini, M. T. M. Khairi: International Review of Automatic Control Vol. 9 (2016), p. 167-204.

[10] S. Sumathi and P. Surekha: Computational Intelligence Paradigms- Theory and Application using MATLAB (CRC Press USA 2010)

IOP Conf. Series: Materials Science and Engineering **303** (2017) 012018 doi:10.1088/1757-899X/303/1/012018

Numerical simulation and experiment on effect of ultrasonic in polymer extrusion processing

Yue Wan[1] ,ZhiHong Fu[1],LingJiao Wei[2],Gongzheng Zang[1] and Lei Zhang[1]

[1]College of Mechanical Electrical Engineering,Central South University,Chang Sha 410083,China
[2]Shanghai Aerospace Science and Technology Institute,China

zhihongfu@csu.edu.cn

Abstract. The influence of ultrasonic wave on the flow field parameters and the precision of extruded products are studied. Firstly, the effect of vibration power on the average velocity of the outlet, the average viscosity of the die section, the average shear rate and the inlet pressure of the die section were studied by using the Polyflow software. Secondly, the effects of ultrasonic strength on the die temperature and the drop of die pressure were studied experimentally by different head temperature and different screw speed. Finally, the relationship between die pressure and extrusion flow rate under different ultrasonic power were studied through experiments.

1. Introduction

Extrusion technology can be molded almost all of the thermoplastic, not only continuous production, but also high production efficiency, so the polymer processing methods occupy a very important position [1,2].

However, the quality of plastic products are affected by the performance of the plastic itself, In the molten state, the macromolecular structure of the polymer chains have a great influence on the stress of the melt viscosity, elasticity and relaxation method. Most linear polymers in the molten state, showed irregular macromolecular chains. Therefore, in the extrusion process, the molecular chain clusters flow through the die when there is extended orientation problem: When the melt flow out of the mold, part of the melt molecular chain elasticity recover, and tend to be line-like, then there will be squeeze swelling phenomenon.

In recent years, as a new energy, ultrasonic energy is often used into conventional processing method [3]. When the intensity is high, it can interact with its acoustic media, change or even destroy the original nature and structure of its acoustic media, this time called power ultrasound [4-6]. At present, domestic and foreign scholars use ultrasonic researches in the field of polymer processing,which are mostly around the ultrasonic impact on polymer melt rheological properties to improve the quality of molding process.[7] When the ultrasonic wave is added to the polymer solution, it can cause the polymer molecular chain to break into macromolecule radicals,and it will change the rheological properties of the polymer melt, and ultimately change the material processing properties and product properties.

In this paper, the rheological behavior of polymer melt under vertical superimposed ultrasonic vibration is studied. Firstly, the effects of ultrasonic wave on the average velocity inlet pressure, average shear rate and the average viscosity of the die were analyzed by numerical simulation. The effect of ultrasonic wave on the head temperature, extrusion pressure and flow rate were studied by ultrasonic assisted extrusion experiment.

2. Simulation

2.1 The Establishment of Geometric Model

Using the capillary die for numerical simulation,circular cross-sectional, the size is shown in Figure 1.

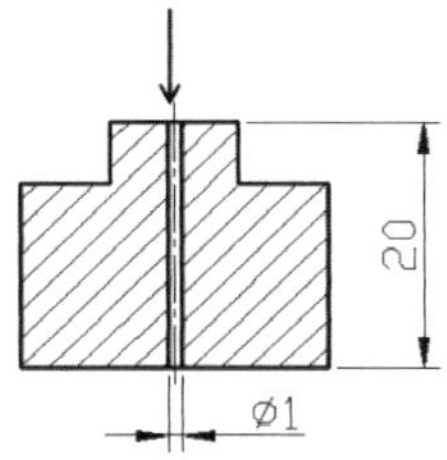

Figure 1. a sectional view of a capillary die model Boundary connection

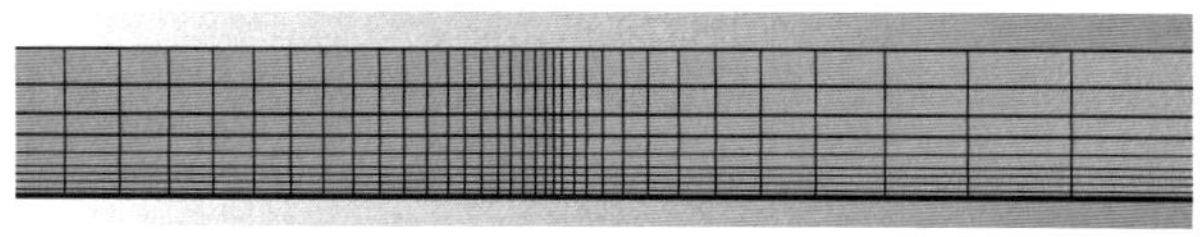

Figure 2. Local enlargement view of finite element

This article uses Ansys own meshing method for structural meshing, and locally refine the die segment and the free segment boundary grid, as shown in Figure 2.

2.2 Mathematical Model

Ultrasonic assisted extrusion process is different from the traditional steady-state extrusion process, which is a cyclical dynamic forming process. In the process of ultrasonic assisted extrusion, due to the introduction of ultrasonic vibration field, The physical parameters (such as inlet flow rate, boundary conditions and material parameters) that characterize the melt flow are time-varying during the melt flow and are transient problems. Combining the basic theory of polymer rheology and the actual situation of melt flow in capillary die, the following assumptions are made[8]:

(1) the melt in the mold will do the full development of laminar motion;
(2) the melt is an incompressible non-Newtonian power law fluid;
(3) ignore the effects of gravity and inertia;
(4) do not consider the effect of wall slip on melt flow

The basic equation for incompressible melt flow:

Continuity equation:

$$\nabla v = 0 \tag{1}$$

Equation of motion:

$$\nabla \tau - \nabla p = 0 \tag{2}$$

Where : v——speed, m/s, t——time, s, p——pressure,Pa ρ——density, kg/m^3, τ_{ij}——Stress deflection, Pa.

The constitutive equation of the melt is Ostwald's power-law constitutive equation, The power-law constitutive equations commonly used in polymer melt processing proposed by Ostwald and de Waele [8] can be used to describe this shear thinning behavior of polymer melts. The power law constitutive equations are as follows:

$$\tau = K\gamma^n \tag{3}$$

Where τ is the shear stress, Pa; K is the melt viscosity coeffient, $Pa \cdot s^n$; γ is the shear rate, s^{-1}; n is the non-Newtonian index.The parameters used in Table 1.

Table 1 .Consistency Coefficient K and Non-Newtonian Index at Different Ultrasonic Power n

Ultrasonic power(W)	Consistency coefficient($Pa.s^n$)	Non - Newtonian index n
0	15217.68	0.35846
300	13216.45	0.4578
500	10015.23	0.5632
800	8996.47	0.6835

2.3. Effect of Vibration Parameters on Flow Field during Ultrasonic Vibration Extrusion

In order to study the effect of ultrasonic vibration parameters on the physical field of polymer melt extrusion, the results of Polyflow can be statistically analyzed.That is,to calculate the mean of the instantaneous value of each parameter in a vibration period. This can be more directly observed the impact of ultrasonic vibration parameters on the physical field. In order to be closer to the actual situation, this section simulates the distribution of the physical fields within 1 minute and processes the results. The simulation results are shown in Fig.3.

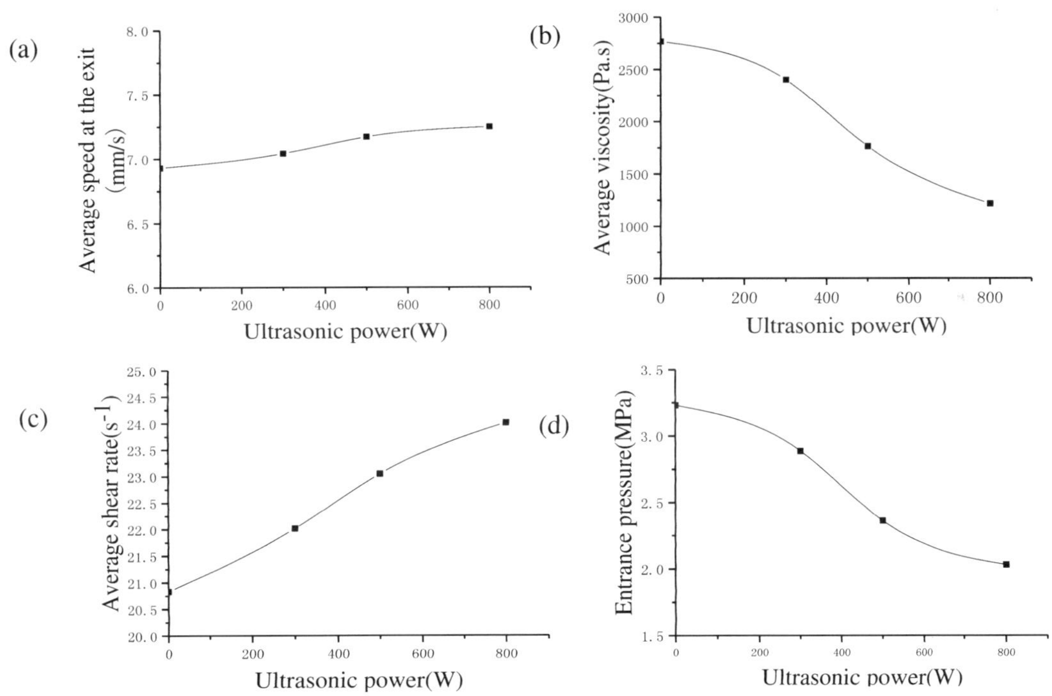

Figure 3 .The relationship between the mean of the parameters of the flow field and the vibration power in one cycle.(a) The average speed at the exit (b) Average viscosity of die module (c) Average shear rate of die (d) Entrance pressure

It can be seen from Fig. (A) that the ultrasonic velocity has little effect on the average velocity of the outlet section when the extrusion volume flow rate is constant during a cycle. Figure (b) shows the average viscosity statistical plot over the entire cross section of the melt in a die during a cycle. It can be seen from the figure that, with the increase of the ultrasonic power, melt viscosity decreased rapidly.

Figure (c) shows the statistical curve of the average shear rate on the entire cross section of the melt in a cycle. It can be seen from the figure that with the increase of the ultrasonic power, the shear rate of the melt in the capillary die increases gradually. Figure (d) shows the average pressure statistics at the inlet of the die during a cycle. With the increase of ultrasonic power, the pressure at the inlet of the die is obviously reduced.

3. Experiment of ultrasonic assisted micro-extrusion

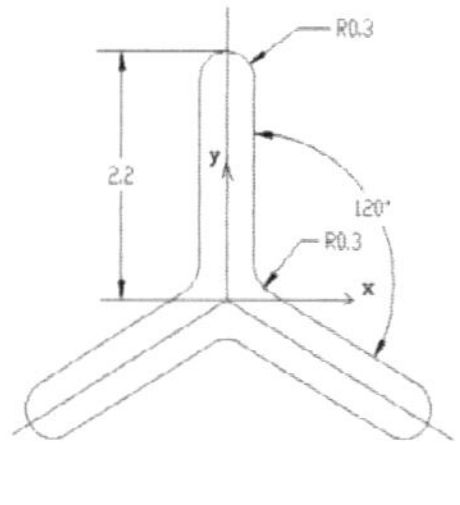

Figure 4. shape and size of die

The low-density polyethylene (LDPE) trilobal microstructure was extruded by ultrasonic - assisted micro - extrusion device. The changes of die temperature, die pressure, extrusion flow rate and dimensional precision of extruder were analyzed after ultrasonic pressure of different power intensity under different die temperature and extrusion speed. Size of trilobal -microstructure die shown in Figure 4.

Experimental materials are the granular low-density polyethylene, South Korea's LG production, model is FB9500. A density of 0.954 g / cm3, a melting point of 112-120 °C, a decomposition temperature of about 300 ° C and a melt index of 3.5 g / 10 min.

In order to test the effect of different processing temperature, screw speed and ultrasonic power on the melt extrusion process, the experimental scheme was designed when the head temperature was 180, 200 and 220 ℃, respectively. Each processing temperature takes three kinds of screw speed at the head of each program, respectively 2r / min, 4r / min and 6r / min. The rheological data of the polymer melt at 0W, 300W, 400W, 500W, 600W and 800W were selected at each screw speed.

4. Effect of extrusion process parameters on forming accuracy

4.1 Effect of ultrasonic wave on extrusion die temperature

Figure 5 shows the effect of ultrasonic intensity on the melt temperature in the die at different experimental temperatures. It can be seen from the figure that when the extrusion temperature is between 180-220 °C, the melt temperature at the outlet of the die is increased with the increase of the ultrasonic power. The lower the head temperature, the greater the effect of ultrasonic on the melt temperature will be in the die. When the head temperature is between 200 °C - 220 °C, the effect of ultrasonic power on the mold temperature is very small.

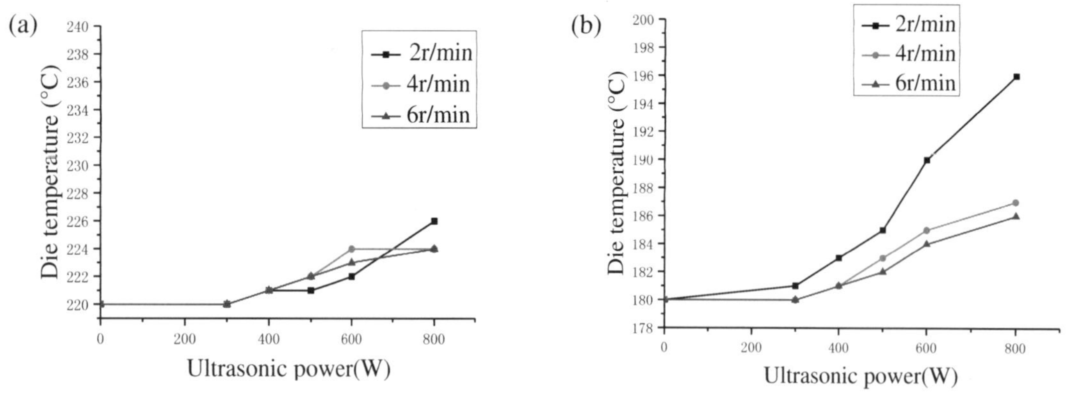

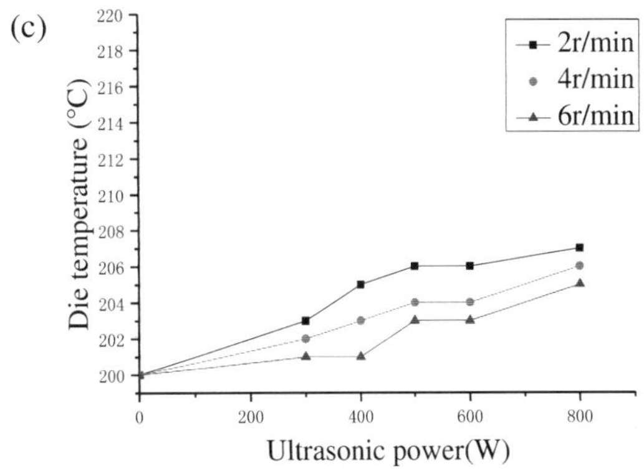

Figure 5. The effect of ultrasonic intensity on the die temperature at different head temperatures.(a) The head temperature is 180 °C (b) The head temperature is 200°C (c) The head temperature is 220°C

4.2 Effects of Ultrasonic on Extrusion Pressure and Flow

Figure 6 shows the effect of ultrasonic intensity on the pressure drop of the extrusion die when the extrusion temperature is constant and the screw speed is different. In the figure, ΔP is used to denote the relative descent value of the die pressure under the action of ultrasonic waves: In the formula, P is the die pressure when no ultrasonic wave is applied, and P_u is the die pressure at the time of applying ultrasonic waves.

$$\Delta P = \frac{P-P_u}{P} \quad (4)$$

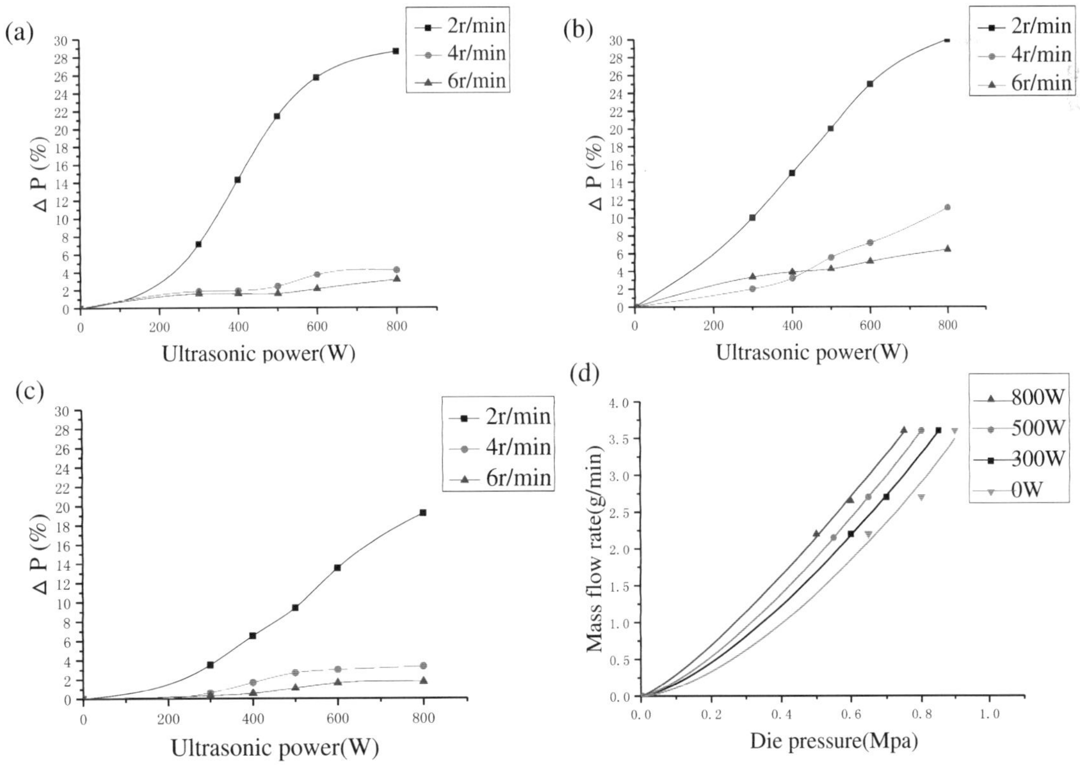

Figure 6 . The effect of ultrasonic intensity on the pressure drop of the die at different head temperatures.(a) The head temperature is 180 °C (b) The head temperature is 200 ° C (c) The head temperature is 220 ° C (d) The relationship between die pressure and extrusion flow

As can be seen from the figure 6,when the extrusion temperature between 180-220 ℃, the ultrasonic can significantly reduces the extrusion die pressure. When the processing temperature is 180 ℃, the screw speed is 2r / min and the ultrasonic power is 800W, the extrusion die pressure drop reaches 29%.

Figure 6 (d)shows the relationship between extrusion pressure extrusion flow ,which is under the extrusion temperature of 180 ℃ and different power of ultrasound. It can be seen from the figure that the extrusion flow increases with the increase of die pressure.

4.3 Effects of Ultrasonic on Extrusion Accuracy

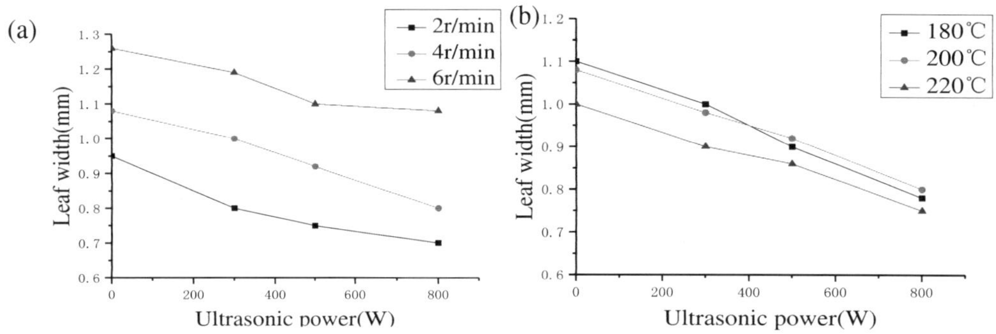

Figure 7. Size of extrudate at different ultrasonic power,(a) The relationship between ultrasonic power and leaf width at 200 °C (b) The relationship between ultrasonic power and leaf width at 4r / min

Figure 7 shows the size of the extrudate at different ultrasonic powers. Figure a) shows the impact of ultrasonic strength on the size of extrudates at different speeds when the head temperature is 200 °C. As can be seen from the figure,as it is with the head temperature of 180°C, the ultrasonic can reduce the leaf width of the trilobical products, that is, to reduce the expansion rate of exports; When the ultrasonic power is constant, the leaf width of the trilobule product increases with the increase of the rotational speed of the screw.

Figure b) shows the effect of ultrasonic intensity on the size of the extrudate at different processing temperatures when the screw speed is 4r / min. It can be seen from the figure that the lower the temperature, the more obvious the effect of ultrasonic on the melt.

5. Conclusion

The simulation results show that the ultrasonic vibration can reduce the inlet pressure of the die, reduce the viscosity and shear stress of the melt in the die, promote the flow of the melt and reduce the resistance of the melt.

Through the experimental study, it was found that the melt temperature at the outlet of the die was increased with the increase of the ultrasonic power. The lower the head temperature, the greater the effect of ultrasonic on the melt temperature in the die. With the increase of the head temperature, the ultrasonic power has little effect on the die temperature. At the same time, the ultrasonic effect can obviously reduce the extrusion die pressure. When the processing temperature is 180 °C, the screw speed is 2r / min and the ultrasonic power is 800W, the extrusion die pressure drop reaches 29%. When the extrusion temperature is 180 °C, the extrusion flow rate increases with the die pressure rising.

Ultrasonic vibration can reduce the extrusion pressure,and can improve the extrusion yield and extrusion efficiency in low extrusion pressure . And ultrasonic vibration assisted extrusion can reduce the export of extruded outlet expansion rate and improve the dimensional accuracy of extruded products.

References

[1] Lin B, Xie Y,Wu M. Research on the micro-extrusion characteristic of mini-screw in the screw extruding spray head. *Polymer Bulletin*, 2010,64(7):727-738.

[2] Sirbu,NA , Serban,VA, Savu,S,in:Influence of the Ultrasonic Microvibrations under the Processing of the HDPE Thermoplastic Material, *MATERIALE PLASTICE*,2014,VOL.51(4)

[3] Cao,Yurong, Li,Huilin,in:Influence of ultrasound on the processing and structure of polypropylene during extrusion, *Polymer Engineering and Science*, 2002, Vol.42 (7), pp.1534-1540

[4] Mekaru H, Noguchi T, Goto H, et al. Effect of applying ultrasonic vibration in thermal nanoimprint lithography. *Microsystem Technologies*, 2008,14(9-11): 1325-1333.

[5] Guo, S., Li, Y., Chen, G. and Li, H., Ultrasonic improvement of rheological and processing behaviour of LLDPE during extrusion. *Polym.* Int., 2003,52: 68–73.

[6] Liu, G. and Li, H., Extrusion of ultrahigh molecular weight polyethylene under ultrasonic vibration field. *J. Appl. Polym. Sci.*, 2003,89: 2628–2632.

[7] Zhao, D., Jin, Y., Wang, M., Li, K., & Song, M. (2009, August). Numerical simulation and experimental study of polymer micro extrusion flow. *International Conference on Mechatronics and Automation2009(ICMA2009)*, (pp. 3155-3160). IEEE.

[8] Hans Jürgen Grieß, Teodor I. Burghelea, Helmut Münstedt ,in: Velocity measurements on a polypropylene melt during extrusion through a flat coat - hanger die, *Polym Eng Sci*, 2012, Vol.52 (3)

Acknowledgments

This work was financially supported by the Hunan Province Key R & D program (application of basic research)(2015JC3008)

IOP Conf. Series: Materials Science and Engineering **303** (2017) 012019 doi:10.1088/1757-899X/303/1/012019

The numerical simulation and experiment on extrusion roller embossing of light diffusion plate with micro-structure

Gongzheng Zang[1], Zhihong Fu[1], Lei Zhang[1], Yue Wan[1]

[1]College of Mechanical Electrical Engineering, Central South University, Changsha 410083, China

zhihongfu@csu.edu.cn

Abstract: Extrusion roller embossing process has demonstrated the ability to produce polymer film with micro-structure. However the influence of various parameters on the forming quality has not been understood clearly. In this paper, a light diffusion plate with semi cylindrical micro-structure array as the research object, the influence of the main processing parameters such as roller speed, pressuring distance and polymer film temperature to the rolling quality was investigated in detail by simulation and experimental methods. The results show that the thickness of the light diffusion plate and the micro-structure fitting diameter increases with the increasing of the roll speed and the polymer film temperature, and decreases with the increasing of the pressing distance. Besides, the simulation results conformed well to the experimental results.

1. Introduction

With the development and application of polymer micro-devices in optics, biology and other fields, micro-nanometer manufacturing technology has gradually become the research focus of related scholars[1]. In 2007, Jiang et al.[2] proposed a novel fabrication method which is based on the traditional micro-hot embossing technology. The film extrusion equipment and R2R hot embossing device were used, and the material is fully warmed and then embossed. This method has the advantages of low forming pressure, high efficiency and continuous production.

Extrusion roller embossing has attracted the interest of scholars since it was proposed. Yeo et al. [3] explored and discussed the effects of the process parameters, such as embossing pressure, substrate preheating, roller temperature and roller speed, on the embossed channel depth, with the goal of achieving process optimisation. Raymond Frenkel et al.[4]designed a variotherm extrusion roller imprinting system. The extruded polymer film was embossed between the belt mold with micro-structure and the pressure roller. Due to the variotherm capability, it was beneficial to reduce the springback of polymer micro-structure after demoulding. Wenlong Liu et al.[5]introduced a new production process. In this process, the thickness of the polymer film was planished and adjusted after it was extruded from the extruder to ensure the thickness and uniformity of the film. In addition. Some people focus on simulation studies. M. Sahli et al.[6] used LSDYNA software to investigate the filling of micro cavities with COC polymer during the roll embossing process and analyse the polymer deformation. Deng Y, et al.[7] established a 3D numerical model of the R2R hot embossing process based on the viscoelastic of PVC polymer to analyse the flow behaviors of the polymer in the filling stage and the demoulding stage.

Although extrusion roller embossing has demonstrated great commercial potential and application prospect in many fields, it still faces many challenges in forming theory, forming process and other aspects. In addition, due to the viscoelasticity of the polymer material, the microstructure on the roller cannot be completely imprinted onto the polymer film. So how to control the extrusion process parameters to obtain the desired size of the product, and the influence of various parameters on the forming quality has not been studied clearly, these problems need to be resolved in order to achieve industrial production.

In this paper, a light diffusion plate with semi cylindrical micro-structure array as the research object, the heat curable material PMMA was employed. Single-factor analysis was adopted to analyze the data. The thickness of the light diffusion plate and the fitting diameter of the micro-structure were taken as the evaluation criterions. Detailed analysis was made to investigate the influence of the main processing parameters such as roller speed, pressuring distance (between substrate roller and drive roller) and polymer film temperature to the rolling quality by the ABAQUS software and the independent building experiment platform. Finally, comparisons made between the experiment and simulation results, and approve the forecast capability of FEM simulation for extrusion roller embossing process.

2. Numerical simulation and experiment

2.1 Experiment

The main process of the roll embossing is illustrated in Figure 1. The semi-circle micro-structure with a radius of 0.5mm is evenly distributed on the substrate roller surface, and the diameter of the two rollers is 65mm. The thickness of PMMA film is 1mm. The material employed in the experiment is Polymethylmethacrylate (VH001) of Japan Mitsubishi Company. Firstly, the PMMA film is extruded from the coat-hanger die. Secondly, it is cooled to a certain temperature, and fed into the rolling device. In this stage, the PMMA film is embossing at a high distance between substrate roller and drive roller, after the embossing process is stable, the parameters are adjusted to the experiment of setting value and begin the experiment.

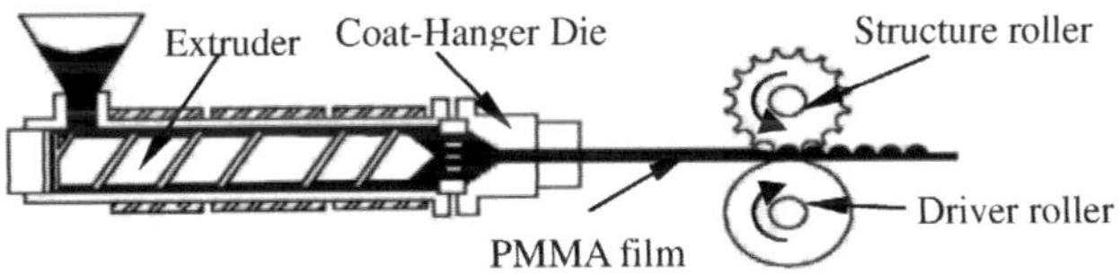

Figure 1. Schematic diagram of the extrusion roller embossing

2.2 Numerical simulation

The viscoelastic model is utilized to simulate the properties of PMMA material. The Wiechert model is adopted to describe the stress relaxation and creep of PMMA, which can be expressed as [8]:

$$E(t) = E_{\infty} + \sum_{i=1}^{n} E_i \exp\left(-\frac{t}{t_i}\right) \tag{1}$$

where E_{∞} is the equilibrium modulus; n is the number of Maxwell units, E_i and λ_i are material constants. And the parameters of the Wiechert model are obtained by uniaxial tensile stress relaxation. Table 1 shows the relaxation modulus and the corresponding relaxation time constant of PMMA material at 105℃. Another the Williams–Landel–Ferry(WLF) equation is used to describe the temperature-dependent modulus of the PMMA material, which can be expressed as[9]:

$$Log_{10}\alpha(T) = -\frac{c_1(T - T_0)}{C_2 + (T - T_0)} \tag{2}$$

where $\alpha(T)$ is the shift factor at temperature T; T_0 is the reference temperature(T_0=105℃); C_1 and C_2 are material constants(C_1=17.6, C_2=65.5). The material parameters used in the simulation are listed in table 2.

Table 1. Parameters of PMMA Wiechert model

n	E_i	λ_i
0	24.495	—
1	20.432	1.039
2	10.685	10.561
3	2.194	94.613

Table 2. Mechanical and thermal parameters of PMMA

Properties	PMMA
Density(g cm^{-3})	1.19
Elastic modulus(MPa)	3300
Poisson's ratio	0.39
Specific heat capacity (J g^{-1}K^{-1})	1.47
Thermal expansion coefficient (K^{-1})	6e^{-5}
Thermal conductivity (W m^{-1}K^{-1})	0.19

The simplified 2D model of extrusion roll embossing was established as shown in Figure 2, and the 1/4 model was selected. Several assumptions were proposed to describe boundary conditions. At first, the bottom of the PMMA film was fixed and the substrate roller was allowed to move downward to pressure the PMMA film. And then, the substrate roller was constrained to move in the -x direction and has a constant rotational velocity about the z direction. Moreover, one considers that the polymer material exhibits the viscoelastic behaviour. The coupled temperature-displacement analysis was considered in considered in the FE model. The roller was treated as an analytical rigid body. The PMMA film was treated as a deformable body and meshed with temperature-displacement coupled elements (CPE4RT) analyzed in plane strain case.

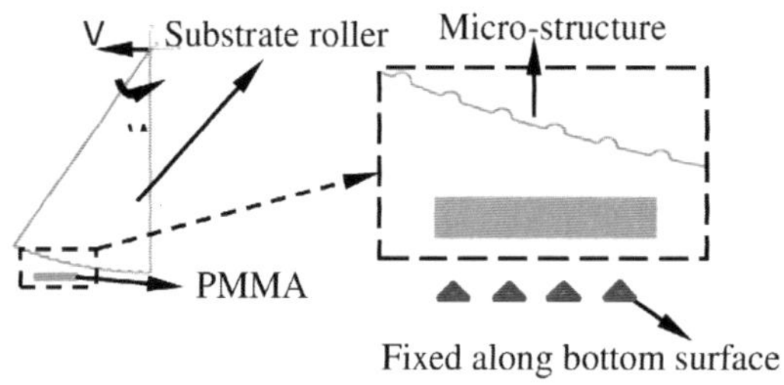

Figure 2. FE model of the extrusion roller embossing process

3. Results and discussions

3.1 Effects of roller speed

The experiment and simulation were conducted at five different roller speeds (1.5rpm, 1.6rpm, 1.7rpm, 1.8rpm and 1.9 rpm), with a 1mm film thickness, pressing distance of 0.7mm and film temperature of 140℃. The light diffusion plate section under the processing conditions of V_{Roll}=1.8rpm is given in Figure 3 (In order to observe, film section of experiment results were blackened. The U_2 in the simulation result is the Y directional displacement).

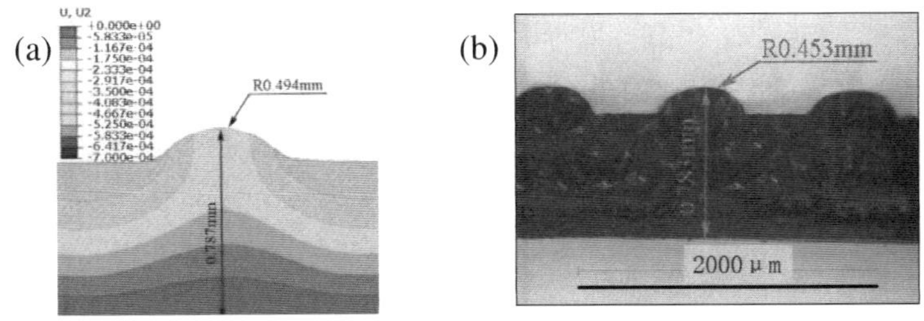

Figure 3. The light diffusion plate section under the processing conditions of V_{Roll}=1.8rpm: (a)simulation result; (b)experience result

The effect of the roller speed on the thickness of the light diffusion plate and the microstructural fitting diameter is shown in Figure 4. According to the Figure 4, with the increasing of the roller speed, the thickness of the light diffusion plate raises gradually, the width and height of micro-structure also raise gradually, so does the fitting diameter. Because if the strain value is constant, the stress will decline with increasing time, which will transform part of the elastic deformation into viscosity deformation, the larger the forming time is, the better the transition is. And the elastic recovery of the micro-structure will be smaller after demoulding. Therefore, when the roller speed increases, the contact time between the roller and the film becomes shorter, resulting in poor replication accuracy and incomplete filling the mold cavity. Again the elastic recovery becomes larger, and the thickness of the light diffusing plate and the fitting diameter of the micro-structure increase.

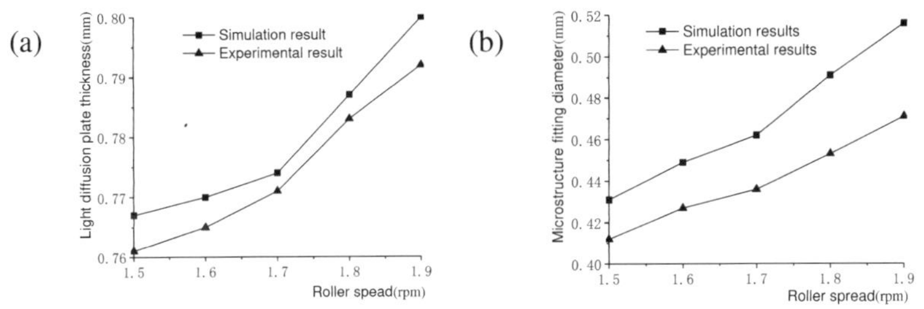

Figure 4. Effect of roller speed on the quality: (a) thickness of light diffusion plate; (b) micro-structure fitting diameter

3.2 Effects of pressuring distance

The experiment and simulation were performed with five different pressuring distances ranging from 0.6 to 0.8mm, with a 1mm film thickness, roller speed of 1.7rpm and film temperature of 140℃. The light diffusion plate section under the processing conditions of L=0.65mm is given in Figure 5.

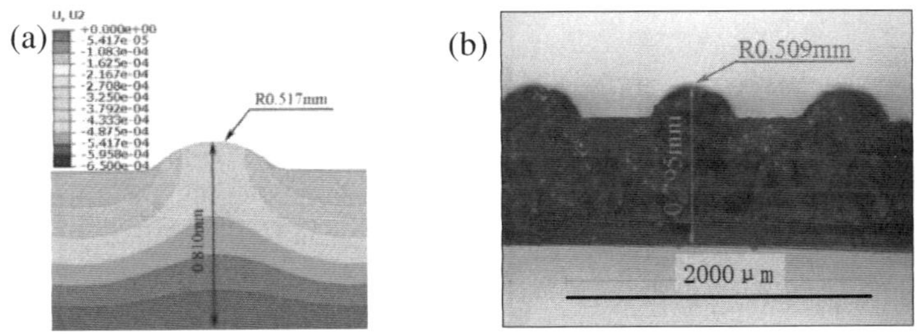

Figure 5. The light diffusion plate section under the processing conditions of L=0.65mm:(a)simulation result; (b)experience result

The effect of pressuring distance on the thickness of the light diffusion plate and the micro-structure fitting diameter is shown in Figure 6. As Figure 6 shows, with the pressing distance increasing, the thickness of the light diffusing plate and the micro-structural fitting diameter gradually decreases. While the height of micro structural increases. The microstructural fitting diameter also decreases little by little. As the increase in pressuring distance is equivalent to increasing the forming pressure, the polymer filling rate gradually increases, mold cavity is very sufficiently filled, and the elastic recovery of the micro-structure is smaller after demoulding. But if the pressure was too high, it will generate residual stress in the products after rolling, and the uneven distribution of the pressure will lead to a poor replication precision.

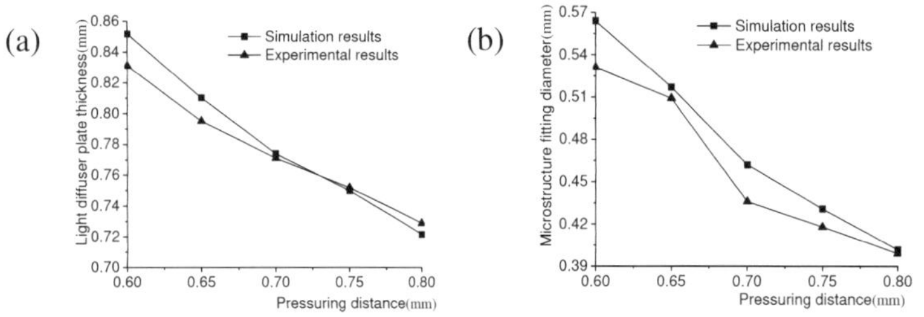

Figure 6. Effect of pressuring distance on the quality: (a) thickness of light diffusion plate;(b) micro-structure fitting diameter

3.3 Effects of film temperature

The experiment and simulation were designed at five different film temperatures (130℃, 135℃, 140℃, 145℃, 150℃) to investigate the forming quality. The pressing distance of 0.7mm and film temperature of 140℃with a 1mm film thickness was used to study the film temperature effects. The light diffusion plate section under the processing conditions of T= 145℃ is given in Figure 7.

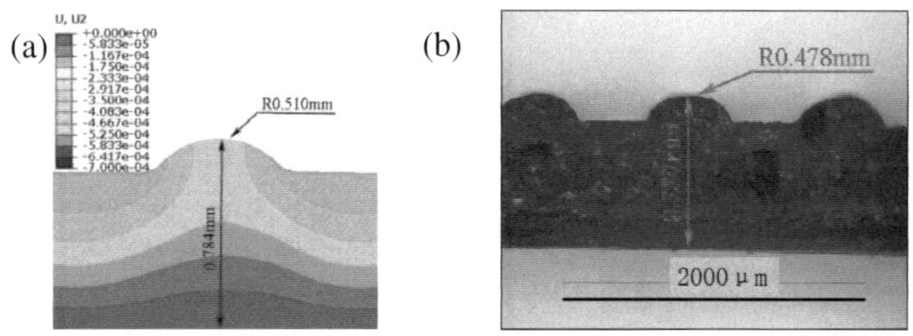

Figure 7. The light diffusion plate section under the processing conditions of T= 145°C:(a)simulation result; (b)experience result

The effect of the film temperature on the thickness of the light diffusion plate and the micro-structural fitting diameter is shown in Figure 8. From Figure 8, the thickness of the light diffusion plate gradually increases with the increasing of the film temperature, the fitting of the microstructure also gradually increases. That is because the film temperature has a significant impact on the flow ability of the

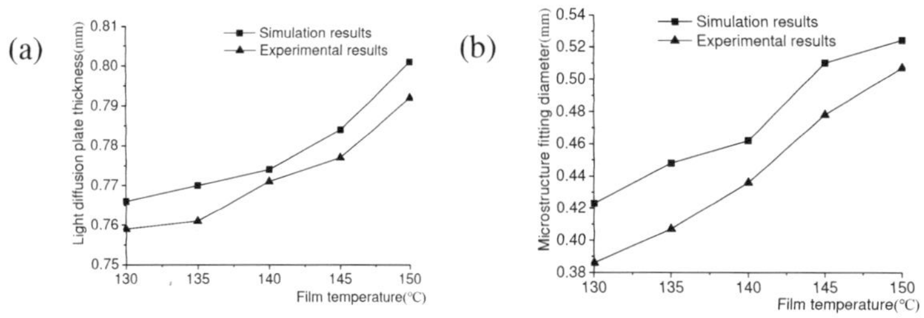

Figure 8. Effects of film temperature on the quality: (a) thickness of light diffusion plate; (b) micro-structure fitting diameter

polymer material. The higher the polymer film temperature is, the better the flow ability is, and the more easier the mold cavity is filled. But if the temperature of the polymer film is too high, the stability of the polymer structure can be destroyed, thus generating many defects on the micro-structure, it also causes residual stress in the final products.

4. Conclusions

In this work, a light diffusion plate was fabricated by extrusion roller embossing process. By means of numerical simulation and experimental investigation, the influence of critical parameters such as roller speed, pressuring distance and polymer film temperature during the extrusion roller embossing process on the thickness of the light diffusion plate and the micro-structure fitting diameter were analyzed. Besides, we compared the simulation and experimental results and well summarized the influence rules of the process parameters. Based on the simulation and experimental results, some preliminary conclusions can be drawn:

(1) The thickness of the light diffusion plate and the micro-structure fitting diameter increases with the increasing of the roll speed and the polymer film temperature, and decreases with the increasing of the pressing distance.

(2) Comparing the simulation and experimental results, the relative errors of the thickness of the light diffusion plate is within 3.2%, and the relative errors of the fitting diameter of microstructure is within 9.2%.The simulation results are satisfactory agreement with the experiment results.

References

[1] Peng L, Deng Y, Yi P. Micro hot embossing of thermoplastic polymers: a review [J]. Journal of Micromechanics & Microengineering, 2013, 24(1):013001.

[2] Jiang L T, Huang T C, Chiu C R. Fabrication of plastic microlens arrays using hybrid extrusion rolling embossing with a metallic cylinder mold fabricated using dry film resist[J]. Optics Express, 2007, 15(19):12088-94.

[3] Yeo L P, Ng S H, Wang Z. Micro-fabrication of polymeric devices using hot roller embossing[J]. Microelectronic Engineering, 2009, 86(4-6):933-936.

[4] Frenkel R, Kim B, Yao D. Extrusion Roller Imprinting with a Variotherm Belt Mold[J]. Machines, 2014, 2(4):299-311.

[5] Liu Y, Liu W, Zhang Y. A Novel Extrusion Microns Embossing Method of Polymer Film[J]. Modern Mechanical Engineering, 2012, 02(2):35-40.

[6] Sahli M, Gelin J C, Barrière T. Numerical Modelling of the Polymers Replication in Micro-Cavities by the Roll Embossing Process[J]. International Journal of Material Forming, 2010, 3(1):607-610.

[7] Deng Y, Yi P, Peng L. Flow behavior of polymers during the roll-to-roll hot embossing process[J]. Journal of Micromechanics & Microengineering, 2015, 25(6).

[8] Park S W, Kim Y R. Fitting Prony-Series Viscoelastic Models with Power-Law Presmoothing[J]. Journal of Materials in Civil Engineering, 2001, 13(1):26-32.

[9] Williams M L, Landel R F, Ferry J D. The Temperature Dependence of Relaxation Mechanisms in Amorphous Polymers and Other Glass-forming Liquids[J]. Journal of the American Chemical Society, 1955, 77(14):3701-3707.

Acknowledgments

This work was financially supported by the Hunan Province Key R & D program (application of basic research)(2015JC3008), Fundamental Research Funds for the Central Universities of Central South University (2017zzts648).

IOP Conf. Series: Materials Science and Engineering **303** (2017) 012020 doi:10.1088/1757-899X/303/1/012020

Die design and process optimization of plastic gear extrusion

Lei Zhang[1], Zhihong Fu[1], Chen Yao[1], Gongzheng Zang[1], Yue Wan[1]

[1]College of Mechanical and Electrical Engineering, Central South University, Changsha, 410083,China

zhihongfu@csu.edu.cn

Abstract. The flow velocity of the melt in the extruder was simulated by using software Polyflow, and the size of the die channel with the best flow uniformity was obtained. The die profile shape is obtained by reverse design. The length of the shaping section is determined by Ansys transient thermal analysis. According to the simulation results, the design and manufacture of extrusion die of plastic gear and vacuum cooling setting were obtained. The influence of the five process parameters on the precision of the plastic gear were studied by the single factor analysis method, such as the die temperature T, the screw speed R, the die spacing S, the vacuum degree M and the hauling speed V. The optimal combination of process parameters was obtained by using the neural network particle swarm optimization algorithm(T = 197.05 °C, R = 9.04rpm, S = 67mm, M = -0.0194MPa). The tooth profile deviation of the extruded plastic gear can reach 9 level of accuracy.

1. Introduction

Gear is one of the most widely used transmission parts in mechanical equipment. Plastic gears are rapidly replacing metal gears in some applications with its low noise, light weight, oxidation resistance, corrosion resistance, and self-lubricating material properties[1]. For example, the power steering of a car is delivered through a set of plastic gears. compared with metal materials, plastics has many advantages in cost, design, processing and performance. The freedom of plastic forming design ensures more efficient gear making. Because it can use plastic to form internal gear, gear sets, worm gears and other products, but it's very difficult for the these products to use metal materials to form in a reasonable price. Plastic gears are generally made by injection molding. However, traditional injection molding technology has the disadvantages of difficult mold making, long molding cycle and low production efficiency. However, for the cylindrical plastic gears with cross-section, the products are in line with the features of extrusion molding process. Moreover, with the development of CAE technology and precision extrusion technology, the extrusion of plastic gear would be possible.

2. Design of extrusion die of plastic gear

As for the shape, plastic gears are similar to pipes and bars, so the design of die can refer to that of pipe and bar [2]. In the extrusion of plastic gears, the inner flow channel and the die is the most important factor affecting the plastics gears. Polymer melt flow in the die is very complex, so the design of the flow channel and die is often achieved by adjusting and correcting by experience to meet the requirements of molding.[3-5]. Because the shape of the plastic gear die is complex and the precision is high, the empirical design method is difficult to meet the requirements. Therefore, this paper used finite software Polyflow to do numerical simulation and inverse extrusion research.[6-9]. Firstly, the optimal flow channel structure was determined. Secondly, the cross-section shape of the die was simulated when the inner rib stereotyping length was determined. Finally, the design of die is completed.

According to the comprehensive analysis of the characteristics of extrusion plastic gear, the straight-through die was selected[10].The parameters of inner flow channel of die are determined by the shape of the shunt cone. The structure of the inner flow channel is shown in Fig.1, the main structural factors that affect the melt flow include the inner rib stereotyping length $L1$, the compression length

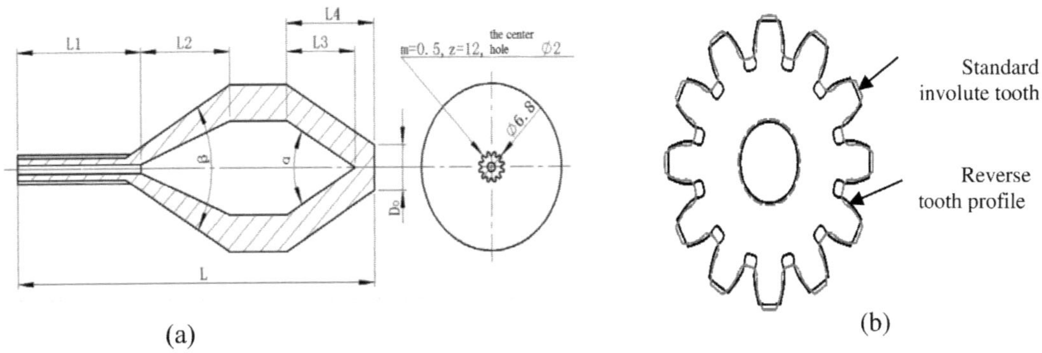

Figure1. (a) The structure of inner flow channel[11] (b) the result of reverse extrusion

$L2$, the shunting cone length $L3$, the shunting segment length $L4$, the total channel length $L4$, the channel inlet diameter D_0, and the compression angle β and the shunting angle α are the main structural factor affecting the melt flow[11].

In this paper, the plastic gear was researched, whose modulus m was 0.5, the number of teeth z was 12, the center hole diameter was 2mm, and the addendum circle diameter was 6.8mm. The optimal flow channel structure parameters were determined by orthogonal test of 16 sets of flow channel parameters by using the software Polyflow. L_1 = 10mm, L_2 = 30mm, L_3 = 23mm, α = 50°, β = 50°, the reverse design of the cross-section of die is shown in Fig. 1. Using Ansys transient thermal analysis, the length of calibrator was 160 mm[12,13].

Straight-through die was used in the extrusion of plastic gear. The profile of die was manufactured by micro EDM technology[14].Vacuum calibrators are used in the extrusion of plastic gear. As shown in Fig.2, the ends of the mold cavity are the cooling room, while the middle is the vacuum chamber[12].

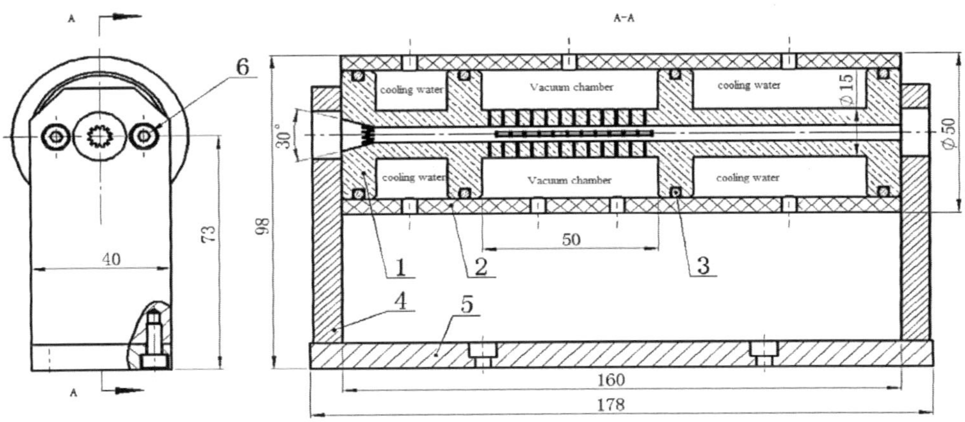

1-shaping tube, 2- plexiglass visual tube, 3- rubber ring, 4- bracket side plate, 5- bracket plate, 6- fastening screw

Figure2. The Vacuum calibrator of plastic gear extrusion

3. Extrusion experiments

The experimental platform mainly consists of the extruder, extrusion die, calibrator, vacuum machine and traction device, as shown in Fig.5. The experimental material is produced by DuPont in America. The model is POM-500P NC010. The material was heated and plasticized by the extruder, the basic shape was completed through the extrusion of die, the plastic gear profile was cooled and shaped by the calibrator, the vacuum machine was to ensure that the calibrator was in a negative pressure state, and finally the formed of plastic gear was pulled out of the calibrator device by the traction device. The characteristic of the experimental platform was that it used the calibrator with cooling function to direct cool, but didn't use the cooling water tank .Therefore it avoided the long distance between the calibrator and traction device, which made the extrusion process unstable and difficult to adjust.

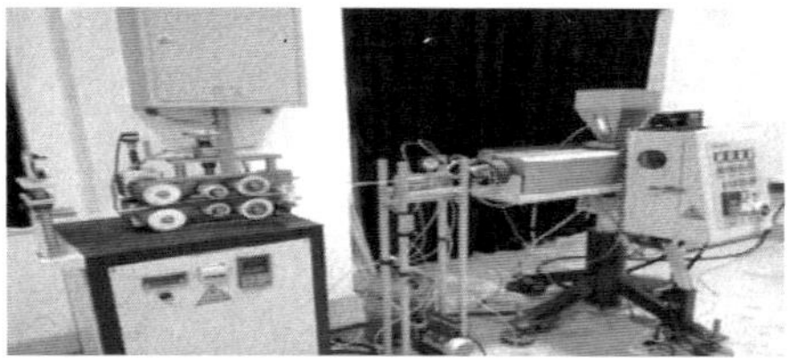

Figure3. the experiment platform for plastic gear extrusion

4. Results and discussion

4.1The analysis of experimental process Parameters

This paper uses the single-factor analysis method to study the effect of five parameters on the forming accuracy of plastic gear, which includes the screw speed, the die temperature, the distance between the die inlet and the die outlet ("die spacing"), the vacuum degree of calibrator and the hauling speed. As the measurement of the accuracy of the extrusion gear was taken by visual parallel detection technology, the four experimental values of the center hole radius, the radius of the tooth circle, the deviation of tooth pitch and the total deviation of tooth profile are taken as the evaluation index of the forming accuracy. The accuracy is different from the general profile of the size requirements of the pipe, the evaluation of more indexes, according to the above four evaluation indexes which are of different importance to accuracy of plastic gears, the four different weight of indexes are to describe the final evaluation of gear forming accuracy F_{ω} , The weight of each evaluation index is shown in Tab.1, and the final evaluation accuracy F_{ω} is calculated as:

$$F_{\omega} = 0.1f_{\varphi} + 0.2f_{\emptyset} + 0.3f_{pt} + 0.4f_{pt}$$

Table1. four corresponding weights of different evaluation indexes

Radius deviation of Center hole f_{φ}	Radius deviation of addendum circle $f_{\emptyset}$	deviation of tooth pitch f_{pt}	Total deviation of tooth profile F_{α}
0.1	0.2	0.3	0.4

The value of the five extrusion process parameters is shown in Tab.2. As shown in Fig.4-7, the influence of each process parameter on the accuracy of forming for plastic gear. The conclusions are as follows: (1) the higher the die temperature is, the higher the accuracy of tooth profile will be. (2) Too small screw speed will affect the fit of the gear profile and calibrator, making the vacuum chamber can't work; the tooth profile accuracy is high when screw speed is between 7 and 9rpm. (3) When the die spacing is too large, the time of gravity action increases, the mobility decreases, and the tooth profile accuracy will be reduced; when the die spacing is too small, it is easy to be sticky. When the die spacing is about 70mm, the tooth shape accuracy is best. (4) The higher the degree of vacuum is, the higher the accuracy of the tooth will be, but high degree of vacuum makes the products and mold friction increase significantly, resulting in hauling difficulties. (5) The higher the hauling speed is, the more the tooth profile accuracy will reduce; high hauling speed will affect the fit of the gear profile and calibrator, making the vacuum chamber can't work properly. (6) The ability of the system to adjust the degree of vacuum would be best when ratio of the screw speed and hauling speed was 4.28.

Table2. extrusion process parameter

Serial number	Die temperature (°C)	Screw speed (rpm)	Die spacing. (cm)	Vacuum degree (MPa)	Hauling speed (mm/s)
1	190	6	60	0	1
2	195	7	70	0.005	1.5
3	200	8	80	0.01	2
4	205	9	90	0.015	2.5
5	210	10	100	0.02	3

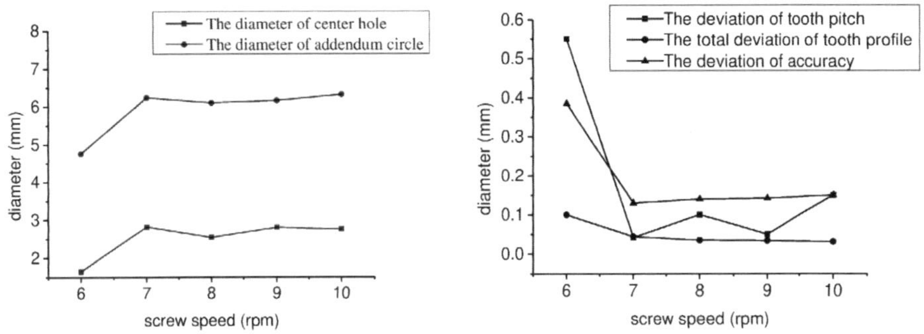

Figure4. The curves between gear cross-section and screwing speed

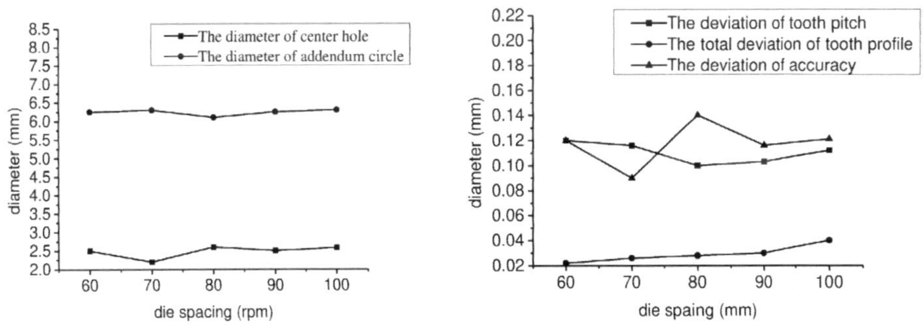

Figure5. The curves between gear cross-section and die spacing

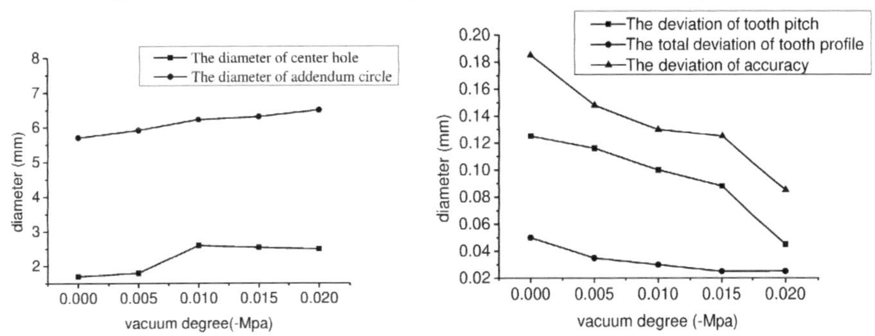

Figure6. The curves between gear cross-section and vacuum degree

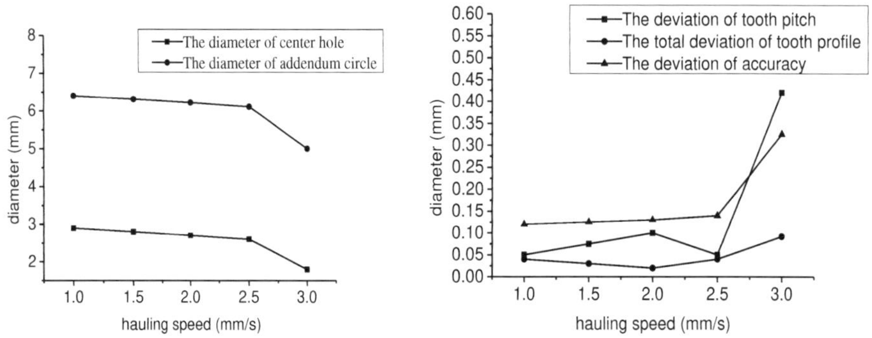

Figure7. The curves between gear cross-section and hauling speed

4.2 Optimization of process parameters for extrusion of plastic gear

Optimization process parameters of extrusion of plastic gear included die temperature T, screw speed R, die spacing S, vacuum degree M, hauling speed V, but it was found that when ratio of the screw speed and the hauling speed was 4.28, the adsorption capacity of the vacuum room would be the best, so we change the process parameters into four, taking the die temperature T, screw speed R, die spacing S and vacuum degree M as the optimization target parameters. For the multi-objective optimization problem, orthogonal test method was generally used in tradition, while extrusion of the plastic gear has higher requirements on the accuracy of the size, and it could only obtain fine combination of process parameters rather than the optimum parameters in the orthogonal test. In this paper, we will firstly select 25 groups parameters to experiment in the way that orthogonal test table select combination of parameters. then conclude the experimental results as BP neural network training learning samples and establish BP network prediction model, finally obtain the sim () function. Then the population was established by IPSO, and the sim () function was used as the fitness function to figure out the optimal combination of process parameters.

The results of orthogonal test were analyzed by extreme difference analysis, and the degree rank of tooth shape accuracy that all process parameters affected were: S> M> T> R. The optimum process parameters were as follows: T = 200 °C, R = 9rpm, S = 70mm, M = -0.02MPa. The process parameters of plastic gear extrusion optimized by BP-IPSO were: T = 197.05 °C, R = 9.04rpm, S = 67mm, M = -0.0194MPa.

The optimum combination from two optimization models was post-rounding experimented according to the actual condition, the corresponding tooth profile accuracy of the indicators are shown in Tab.3. The results show that after using neural network and improved particle swarm optimization algorithm, the optimum accuracy of the extrusion of plastic gear which was 0.0415 in the orthogonal table increased by 27.07%. And the deviation of tooth profile F_{α} which was 0.014 reached 9th level of gear accuracy. It was verified that the BP-IPSO method was effective and reliable in improving the accuracy of gear extrusion. The corresponding extrusion gear samples under the optimum process parameters are shown in Fig.8.

Table3. The accuracy of gear extrusion of two optimization models

	T °C	R rpm	S cm	M -MPa	f_{φ}	$f_{\emptyset}$	f_{pt}	F_{α}	F_{ω}
Orthogonal Optimum	200	9	70	0.02	0.05	0.11	0.027	0.016	0.0415
BP-IPSO Optimum	197.05	9.04	67	0.0194	0.025	0.12	0.003	0.014	0.0326

(a) (b

Figure8. Samples of gear extrusion under optimum process parameters (a) Samples of gear extrusion under orthogonal test. (b) Samples of gear extrusion under BP-IPSO.

5. Conclusion

In this paper, the finite element software was used to simulate the design of extrusion die and vacuum calibrators of plastic gears. The optimum size of the die was determined, which provided the basis for the design and processing. The process parameters that influenced the accuracy of tooth profile were performed by univariate analysis and rules were obtained. The higher the degree of vacuum and die temperature is, the higher the accuracy of the tooth will be, when screw speed is between 7 and 9rpm and the die spacing is about 70mm, the tooth shape accuracy is high. The ability of the system to adjust the degree of vacuum would be best when ratio of the screw speed and hauling speed was 4.28. The parameters of extrusion process were optimized by using the comparison between the orthogonal test and the neural network particle swarm algorithm. T = 197.05 °C, R = 9.04rpm, S = 67mm and M = -0.0194MPa. And the tooth profile deviation of extrusion plastic gear reached 9th level of accuracy which reaches general level. The production efficiency was 10 times higher than that of injection molding. The research in this paper can guide the actual extrusion of plastic gears.

References

[1] Tsai M H, Tsai Y C. A method for calculating static transmission errors of plastic spur gears usingFEM evaluation[J]. Finite Elements in Analysis and Design.27, 345-357(1997).
[2] Hurez P, Tanguy P A, Blouin D. A new design procedure for profile extrusion dies[J]. Polymer Engineering & Science.36,626-635(1996).
[3] MA X, HUANG M. Optimization Design for Profile Die Based on Numerical Simulation [J]. China Plastics Industry.4, 009(2008)..
[4] Wang Q B, Wen X Q, Liu X, et al. Optimization design on extrusion die Channels for Plastic Profile based on the CAD/CAE/ERP [J]. China Plastics Industry.37,26-28(2009).
[5] Yang J L, Liu H S, Huang X Y, et al. Application of Polyflow inverse extrusion method in the design of T-shaped plastics profile extrusion die[J]. Journal of Nanchang University, 2010, 40,114-108(2010).
[6] Pepliński K, Mozer A. Design of extrusion die for plastic profile using Ansys Polyflow software[J]. Journal of Polish CIMAC.6,221-226(2011).
[7] Szarvasy I, Sienz J, Pittman J F T, et al. Computer aided optimisation of profile extrusion dies: definition and assessment of the objective function[J]. International Polymer Processing, 2000, 15,28-39(2000).
[8] Lebaal N, Schmidt F, Puissant S. Design and optimization of three-dimensional extrusion dies, using constraint optimization algorithm[J].Finite Elements in Analysis and Design. 45,333-340(2009).
[9] Liu B, Ma J. Application of polyflow inverse extrusion in the design of profile die[J]. Plastics

Science & Technology.2008.
[10] Liang R F, Mackley M R. The gas-assisted extrusion of molten polyethylene[J]. Journal of Rheology (1978-present).45,211-226(2011).
[11] Fu Z, Yao C, Zhang J, et al. Simulation of Extrusion and Die Design of Plastic Gear[C]// International Conference on Electromechanical Control Technology and Transportation. 2015.
[12] Wang Q B, Wang Z M, Peng A H. Optimization Design of Cooling System of Vacuum Calibrator for Plastics Profile Extrusion[J]. Advanced Materials Research.168-170,959-962(2010).
[13] Zhao D Y, Wang M J, Song M C. Optimization design in cooling channels of vacuum calibrator for plastic profile extrusion[J]. Journal of Dalian University of Technology.47,533-538(2007).
[14] Jin G B, Wang M J, Zhao D Y, et al. Design and experiments of extrusion die for polypropylene five-lumen micro tube[J]. Journal of Materials Processing Technology.214, 50-59(2014).

Acknowledgements
This work was financially supported by Fundamental Research Funds for the Central Universities of C entral South University (2017zzts643). the Hunan Province Key R & D program (application of basic research)(2015JC3008).

The Iron Removal in Marmatite Concentrate Pressure Leaching Process

LUO Wen-bo[a], WANG Ji-kun[a,b], GAN Yin[c]

[a] Faculty of Metallurgical and Energy Engineering, Kunming University of Science and Technology, Kunming, Yunnan, 650093, China
[b] Yunnan Metallurgy Group Parent Company, Kunming, Yunnan 650031, China
[c] Yunan Gold&Mining Group CO.,LTD, Kunming, Yunnan 650224, China

luowenbo@126.com

Abstract. To modify the pressure leaching technology of horizontal autoclave using marmatite concentrate, an appropriate increase in the pulp's residence time in the horizontal autoclave is required. This increase will provide sufficient time for leaching to be completed in the first three chambers of the horizontal autoclave. Adding zinc oxide ore and potassium sulfate in the fourth chamber of the horizontal autoclave is needed to complete preliminary neutralization and iron precipitation in the horizontal autoclave. The pilot plant experimental results of the proposed technology are satisfactory, further shortening the process of pressure leaching and improving its economic efficiency.

1. Introduction

Currently available mineral resources are mostly low-grade and complex with multi-metal mixing because of the depletion of high-quality mineral resources [1]. Marmatite concentrate is a zinc mineral that is categorized into the low-grade and complex class of resources. The iron element in marmatite is embedded in the sphalerite crystal through isomorphous substitution. As a result, general beneficiation methods cannot effectively separate zinc from iron in marmatite. Proven reserves of this mineral in Yunnan province, China, are as high as 7 million tons and accounts for one-third of the proven zinc resource reserves in the province [2, 3].

The traditional zinc hydrometallurgy process requires the iron content to be as low as possible, because iron-containing zinc concentrate generates copious zinc ferrite during roasting, which reduces the efficiency of zinc leaching. Although utilizing hot acid leaching can destroy zinc ferrite and increase the efficiency of zinc leaching, iron is also is leached into the solution. The subsequent removal of iron is a lengthy process that results in a larger amount of leach residue, low zinc recovery rate, and other limitations [4-6]. Using a single-stage pressure leaching technology results in approximately 60% precipitation of the iron into the residues, although the iron content in the leachate can be as high as 5 ~ 15 g/L and the acid concentration is generally 30 g/L or higher [2,3]. The resulting leachate must undergo the jarosite process to remove the irons. In a two-stage pressure leaching technology of marmatite concentrate, the irons and acids contents of the leachate are low, although the iron content is still 3 ~ 5 g/L, which does not meet the requirements of neutral supernatant. Additionally, the capital investment required for the two-stage pressure leaching is very substantial [7]. In the current study, the removal of iron from marmatite via single-stage horizontal autoclave pressure leaching is described. This process removes iron from the leaching solution while maintaining the requirements for a neutral supernatant.

2. Experimental

2.1. Raw Materials
The concentrated sulfuric acid and potassium sulfate used in experiment were industrial grade, the chemical composition of single-stage pressure leaching liquor was shown in Table 1, chemical composition of zinc oxide ore used for neutralization was shown in Table 2.

Table 1. Chemical composition of single-stage pressure leaching liquor.

Component	Zn^{2+}	Fe^{3+}	Fe^{2+}	H_2SO_4	As^{3+}	Cd2+	Co^{2+}	Ni^{2+}	Cu^{2+}
Content(g/L)	128.37	13.21	2.93	42.58	0.087	0.382	0.014	0.022	0.115

Table 2. Chemical composition of zinc oxide ore (mass fraction, %).

Component	Zn	Pb	Fe	CaO	SiO2	As	F	Cl
Content(%,ω)	19.64	3.45	8.51	7.13	10.087	0.18	0.027	0.006

2.2. Experiment Programs
The horizontal autoclave has four chambers. In the experiments, the autoclave system reduced the flow rate of slurry. Thus, the marmatite concentrate was completely leached in the first three chambers. In the fourth chamber, zinc oxide ore and potassium sulfate were added. Under high temperature, the zinc ore and potassium sulfate inside the horizontal autoclave resulted in a high-pressure and oxidizing atmosphere. The iron was quickly precipitated in the form of jarosite salt [9-11]. As a result of leaching, the neutralization and iron removal of the marmatite concentrate were completed in a horizontal autoclave. The experiment process flow chart is presented in Figure 1.

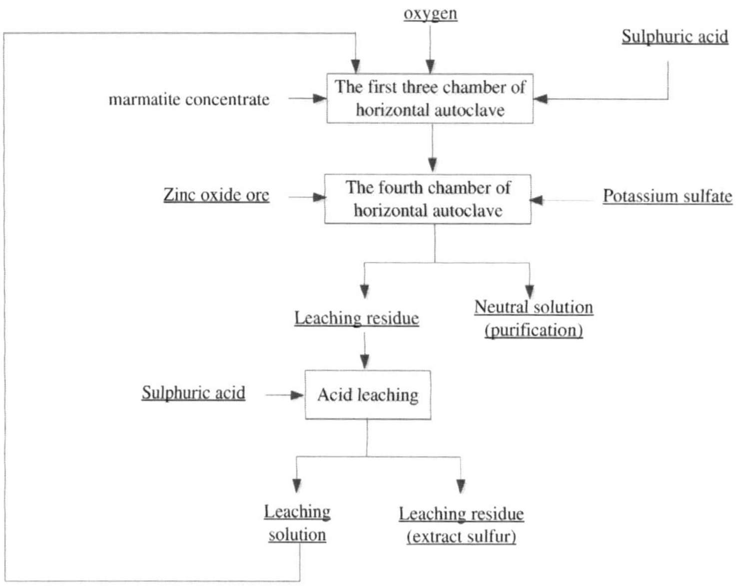

Figure 1. Process chart

IOP Conf. Series: Materials Science and Engineering **303** (2017) 012021 doi:10.1088/1757-899X/303/1/012021

The pressure leaching process of marmatite concentrate is a routine technology, and thus the process will no longer be described here. In these experiments, a single-stage pressure leachate of marmatite concentrate was used as raw materials. The experiment to remove iron was conducted in the 2 L autoclave. The influence of various reaction parameters on iron removal was investigated through a series of carefully controlled experiments. The parameters included final PH value, reaction time, potassium sulfate dosage, temperature, and other factors.

3. Results and Discussion

3.1. Effects of Final pH Value

Through the process of pre-grinding zinc oxide ore the size of particles less than 0.074 mm accounted for 95%. Experiment conditions: pressure leaching liquor = 500 mL, potassium sulfate = 3g, temperature = 130 °C, reaction time = 30 min, air pressure= 1.2 MPa, stirring speed = 600 rpm. The results are shown in Figure 2.

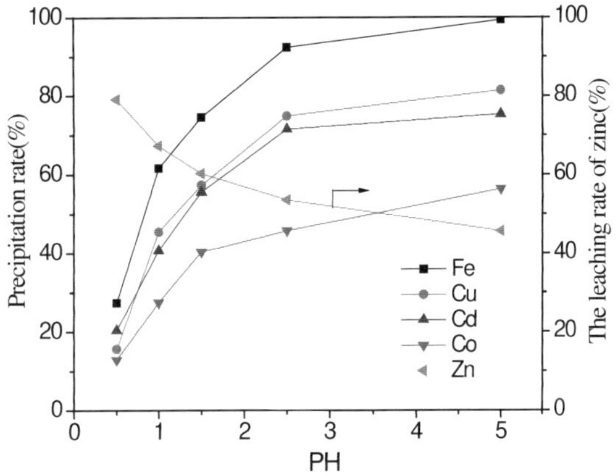

Figure 2. Effects of final pH value

The figure indicates that the precipitation rate of irons and the main impurities of copper, cadmium, and cobalt increased, and the leaching rate of zinc decline with the final pH value of the reaction. This result occurred because the precipitation rate of irons increased with the final pH value, whereas, the quantity of impurities of copper, cadmium, and cobalt increased in the leach residue as a result of hydrolysis precipitation or adsorbed in the iron precipitation. Simultaneously, the filtration rate of slurry decreased with the increase in pH value. To ensure the highest precipitation rate of iron, the final pH value of the slurry should be 5.0.

3.2. Effects of Temperature

Experiment conditions: pressure leaching liquor = 500mL, potassium sulfate = 3g, final pH value = 5.0, reaction time = 30 min, air pressure = 1.2 MPa, stirring speed = 600 rpm. The results are shown in Fig. 3.

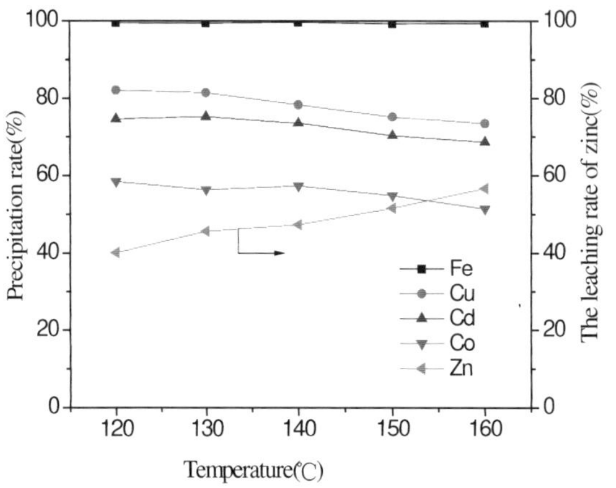

Figure 3. Effect of temperature

The figure shows that the precipitation rate of iron increased with the temperature of the reaction, whereas the impurities of copper, cadmium, and cobalt gradually decrease, and the leaching rate of zinc increases slowly. The reason for this outcome was the leaching residue at high temperature has better crystal shape. Thus, as temperature increased, the amount of absorbed metal impurities decreased in the iron precipitation residue. At the same time, the filtration rate of the slurry increased with temperature. Thus, the optimum operating temperature for this reaction should be near 150 °C.

3.3. Effects of Potassium Sulfate Addition

Experiment conditions: pressure leaching liquor = 500 mL, reaction temperature =150 °C, final pH value = 5.0, reaction time = 30 min, air pressure = 1.2 MPa, stirring speed = 600 rpm. The results are shown in Fig. 4.

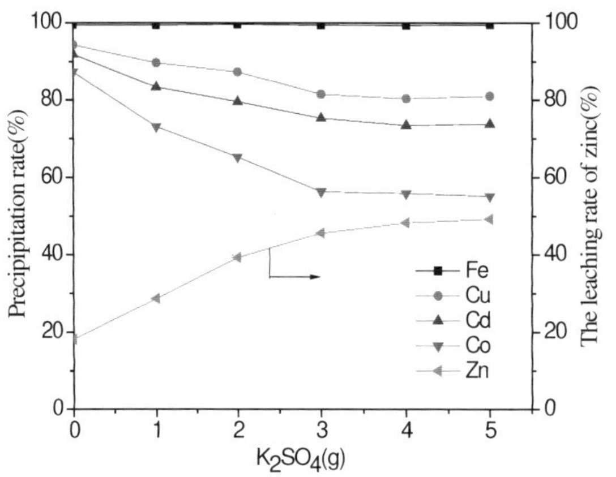

Figure 4. Effects of potassium sulfate addition

The figure indicates that the precipitation rate of iron was essentially the same as that with the increased amount of the potassium sulfate to the reaction mixture. In comparison, the precipitation rate of copper, cadmium, cobalt, and other impurities decline at first and remained constant after, the leaching rate of zinc increase at first and constant after. The leaching residue has better crystal shape and the filtration rate of the slurry increased with the amount of potassium sulfate added. Thus, the amount of potassium sulfate added 3g is appropriate.

3.4. Effects of the Reaction Time

Horizontal autoclave has four reaction chambers, sphalerite pressure leaching time is generally 1 ~ 2.5 h, so each chamber of the residence time is 15 ~ 40 min, due to the iron precipitating process performed in the fourth chamber, therefore the investigated time range is 15 ~ 40 min. Experiment conditions: pressure leaching liquor = 500 mL, reaction temperature = 150 °C, final PH value = 5.0, potassium sulfate = 3 g, air pressure = 1.2 MPa, stirring speed = 600 rpm. The results are shown in Fig. 5.

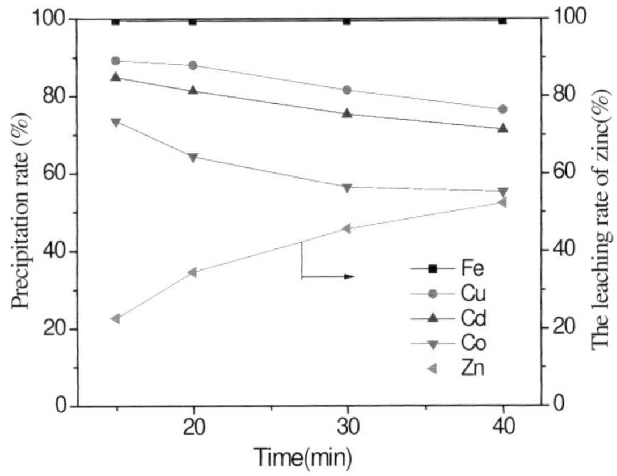

Figure 5. Effects of the reaction time

The figure indicates that the precipitation rate of iron is essentially the same as that with the increased pressure time. In comparison, the precipitation rate of copper, cadmium, cobalt, and other impurities slowly decline, the leaching rate of zinc slowly increase. Thus, the iron removal time should be longer, 30 ~ 40 min is appropriate.

3.5. Horizontal Autoclave Pilot Plant Experiment

The precipitating iron experiment conditions were chosen as follows: temperature = 150 °C, final pH = 5.0, pressure = 1.2 MPa, reaction time = 30 min, and stirring speed = 600 rpm, potassium sulfate = 3 g. The experimental results are presented in Table 3.

Table 3. Chemical composition of leachate from horizontal autoclave.

Order	Composition of leachate (g/L)						
number	Zn	Fe	Cu	Cd	Co	Ni	As
1	142.53	0.134	0.032	0.064	0.006	0.013	0.016
2	148.65	0.107	0.042	0.066	0.005	0.009	0.013
3	141.24	0.158	0.033	0.072	0.006	0.011	0.019
4	152.69	0.132	0.045	0.076	0.008	0.014	0.023

The chemical composition of the leachate from horizontal autoclave pressure leaching in Table 3 demonstrates that the content of harmful impurities in the leachate was relatively low. Thus, the product could be sent directly to the zinc-leaching solution purification process.

4. Conclusions

(1) The optimum conditions of the pressure leaching iron removal were as follow: reaction temperature = 150 °C, end-point PH = 5.0, potassium sulfate = 3 g, pressure = 1.2 MPa, the reaction time = 30 min, stirring speed = 600 rpm. The iron removal process was performed in the fourth chamber of the horizontal autoclave at high temperature and high pressure under strong oxidizing circumstance within the autoclave. Just 30 min to complete the process, and the iron removal effect is obvious.

(2) Compared with the traditional zinc pressure leaching process, the proposed method disregards the jarosite iron-removal process outside the horizontal autoclave steps, shortens the process, and reduces the amount of zinc powder in zinc-leaching solution purification, because of the lower contents of harmful impurities in the leaching solutions. The economic benefits of the proposed process are significant.

References

[1] K Q Xie, X W Yang, Y Z Shu and Q F Shen 2006 *China Nonferrous Metallurgy* vol 4, p 19
[2] J K Wang, T X Zhou and J M Wu 2004 *Nonferrous Metals (Extractive Metallurgy)* vol 1, p 5
[3] J K Wang and T. X. Zhou 2005 *Technology and industrialization of pressure acid leaching on zinc sulfide concentrate* (Beijing: Metallurgical Industry Press) p 45
[4] S C Wang and Y Zhang 2007 *Nonferrous Metallurgy* vol 2 p 19
[5] F C Wei 2008 *China Nonferrous Metallurgy* vol 1 p 21
[6] F Elgersma, G J Witkamp and G M Van Rosmalen 1993 *Hydrometallurgy* vol 33 p 313
[7] F Habashi 1997 *Proc XX IMPC (Aachen)* vol 4 p 129
[8] X G Zhou 1998 *Metallurgical Collections* vol 6 p 18
[9] S G Zhou, L X Zhou and H Z Huang 2004 *Spectroscopy and Spectral Analysis* vol 24 p 1140
[10] J Babcan 1971 *Geol. Zb. (Bratislava)* vol 22 p 299
[11] C Q Wang, S F Ma, A H Lu and J G Zhou 2005 *Acta Petrologica Et Mineralogica* vol 6 p 607

Acknowledgements
Here extend author sincere gratitude to Hexing Company, Jianshui, Yunnan, for their help provided during the tests.

Study on copper kinetics in processing sulphide ore mixed with copper and zinc with sulfuric acid leaching under pressure

LUO Wen-bo[a], WANG Ji-kun[a,b], GAN Yin[c]

[a]Faculty of Metallurgical and Energy Engineering, Kunming University of Science and Technology, Kunming, Yunnan 650093, China
[b]Yunnan Metallurgy Group Parent Company, Kunming, Yunnan 650031, China
[c]Yunan Gold&Mining Group CO.,LTD, Kunming, Yunnan 650224, China

luowenbo@126.com

Abstract. Sulphide ore mixed with copper and zinc is processed with pressure acid leaching. Research is conducted on the copper kinetic. The stirring rate is set at 600 rpm which could eliminate the influence of external diffusions. Research is conducted on the factors affecting the copper leaching kinetic are temperature, pressure, concentration of sulfuric acid, particle size. The result shows that the apparent activity energy is 50.7 KJ/mol. We could determine that the copper leaching process is shrinking core model of chemical reaction control and work out the leaching equation.

1. Introduction

So far, 85% copper around the world is extracted through pyrogenic process and only 15% copper is produced through wet process. But the pyrogenic process requires that the copper content should be high and the detrimental impurity content should be low. Due to the mining of ore resources and the depletion of high-quality mineral resources, most of the resources remained are lean ore and mixed ore[1]. The mixed ore is separated through benefication, not only the metal recovery rate is low but also the content of each metal is high in the concentrate. So it is hard to meet the request of smelting. Pressure leaching is a clean and effective new metallurgy technique which has been applied to zinc sulfide ore, nickel and cobalt extraction, Gold pretreatment, etc as well as the pressure leaching of mixed ore. The results are all quite satisfying[2-4].

So far, the researches on copper leaching kinetic are: cooper leaching kinetic in ferric sulfate acid solution conducted by Misra[5], the copper leaching kinetic in chlorination system conducted by Havlik[6], the copper leaching kinetic in Hydrogen peroxide - ethylene glycol system conducted by Mahajan[7], the low grade copper oxide ore leaching kinetic in Ammonia - ammonium sulphate system conducted by Liu Zhixiong[8], the chalcopyrite dissolution kinetics conducted by Xing Weiguo[9], the pressure copper leaching kinetic in chalcopyrite concentrate by McDonld[10], the pressure copper leaching kinetic in lead matte conducted by Jin Bingjie[11], etc[12-14]. it can be seen that researches on copper leaching kinetic are mostly concentrated on leaching process under normal pressure and there are fewer researches on leaching process under intense pressure.

For making clear the mineral containing copper leaching behavior in processing sulphide ore mixed with copper and zinc with sulfuric acid leaching under pressure and facilitate the leaching of copper, this paper conducts research on copper leaching kinetic in sulphide ore mixed with copper and zinc to find out the main factors affecting copper indium leaching kinetic and find out the reaction controlling steps and set up the kinetic equation.

2. Test materials and methods.

2.1 test materials

The test materials are sulphide ore mixed with copper and zinc obtained in Yunnan Province, China. The chemical composition of the sample is shown in Table 1.

Table1. The main components

Component	Zn	Cu	Fe	S	Pb	Ag	Au
Content (%)	28.65	5.22	10.95	14.23	0.38	0.0389	0.0044

The materials are sulphide ore containing copper, zinc, iron and a part of oxidized ore. Analysis on the chemical matter phase shows that zinc, copper and iron exist in the form of ZnS, $CuFeS_2$ and $Fe_{1-x}S$.

2.2 Testing method

Mixed ore and Sulfuric acid leaching agent are filled in 2 L autoclave according to a certain proportion and then seal the autoclave. Fill in the oxygen till the pressure reach the set value and increase the temperature at the same time, start stirring until the temperature reaches the set value and start timing, keep stirring while keeping the pressure and temperature inside the autoclave at the set value until the set time. After the reaction is completed, fill the cool water in the autoclave to reduce the temperature to 90 °C. Release the pressure and stop stirring. Open the autoclave cover and take out the pulp and filter it. Take the sample of filtrate and leaching residues to analyze. To make sure the concentration of leaching agent is steady, we set the liquid-solid ratio at 50:1. That is to say 10 g mineral and 500 mL sulfuric acid leaching agent are used in every test and keep the rate of temperature increase and decrease at around the same value.

3. Result and discussion

The main chemical reaction happened in the pressure leaching process:

$$CuFeS_2 + 2H_2SO_4 + O_2 = FeSO_4 + CuSO_4 + 2S + 2H_2 \quad (1)$$

$$Fe_{1-x}S + (1-x)H_2SO_4 + (1-x)/2O_2 = (1-x)FeSO_4 + S + (1-x)H_2O \quad (2)$$

$$ZnS + H_2SO_4 + 0.5O_2 = ZnSO_4 + S + H_2O \quad (3)$$

The above test (1) and (3) show that there is solid product Sulphur produced. We have reason to believe that this liquid-solid reaction process is shrinking core model with solid products generated. The reaction steps of shrinking core model include: 1). the liquid reactant diffuse externally through the liquid boundary layer. 2) The liquid reactant diffuses internally through the solid product layer. 3) Interface chemical reaction. The leaching rate is decided by the component with the lowest reaction rate[15-16].

The kinetic equation when the reaction is controlled by external diffusions reaction [15]:

$$\alpha = k_c t \quad (4)$$

where α—— the leaching rate of solid reactant;

k_c——the rate constant of external diffusions;

t——time, min;

The kinetic equation when the reaction is controlled by internal diffusions reaction [15]:

$$1-2/3\alpha-\left(1-\alpha\right)^{2/3}=k_d t \quad (5)$$

where k_d——the rate constant of internal diffusions;

The kinetic equation when the reaction is controlled by chemical reaction [15]:

$$1-\left(1-\alpha\right)^{1/3}=k_r t \quad (6)$$

where k_r——the rate constant of chemical reaction;

We could get the copper leaching kinetic equation when we work out the speed constant K'.

3.1. The effect of stirring rate on copper leaching

When the particle size is 80~96 μm, the temperature is 130 °C, the liquid-solid ratio is 50 mL/g, the concentration of sulfuric acid is 50 g/L, the pressure is 1.0 Mpa, the time is 1 h, a small amount of Surfactant lignin is added, the effect of stirring rate on the copper leaching rate is shown in Fig. 1.

Fig. 1. shows that when the stirring speed is over 600 rpm, it has no effect on the rate of copper leaching. It means that when the stirring rate is over 600 rpm, therefore, in the future test, we set the stirring speed at 600 rpm to eliminate the effect of stirring rate on copper leaching.

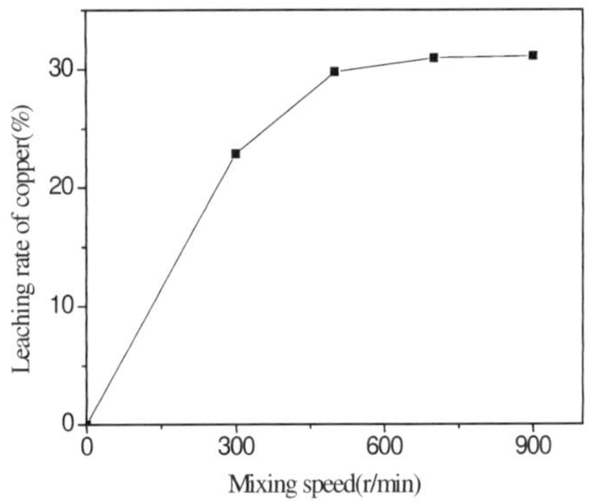

Figure.1. Effect of mixing speed on leaching rate of copper

3.2 The effect of temperature on copper leaching rate

When the particle size is 80~96 μm, the concentration of sulfuric acid is 50 g/L, the pressure is 1.0 Mpa, the stirring rate is 600 rpm, a small amount of surfactant lignin is added, the effect of temperature on the copper leaching rate is shown in Fig. 2.

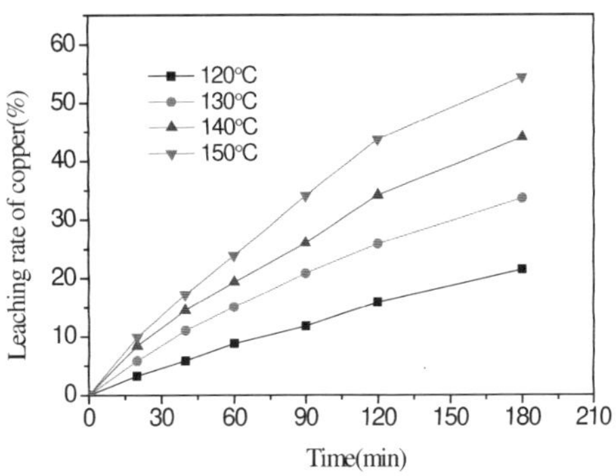

Figure. 2. Effect of temperature

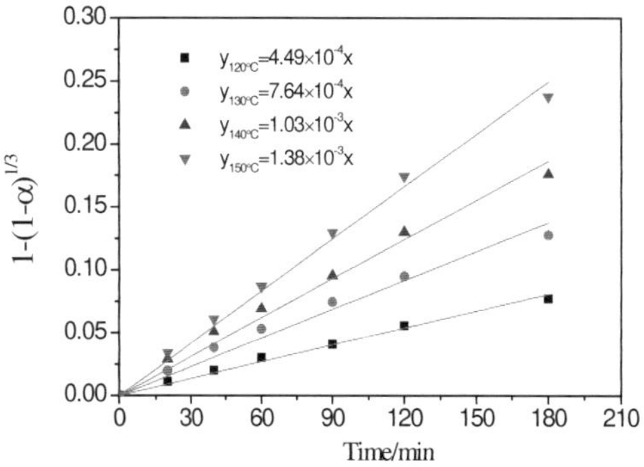

Figure. 3. Relationship of temperature $1-(1-\alpha)^{1/3}$ with time

We could work out the relationship of $1-(1-\alpha)^{1/3}$ and $1-2\alpha/3-(1-\alpha)^{2/3}$ with time. The result shows in Fig. 3 and 4.

From Fig. 3 and Fig. 4, we could see the coefficient all bigger than 0.991 and the line fitting is good. Fig. 4 shows that the smallest coefficient is 0.935 and the line fitting is bad. The preliminary result shows that the leaching process is controlled by chemical reaction.

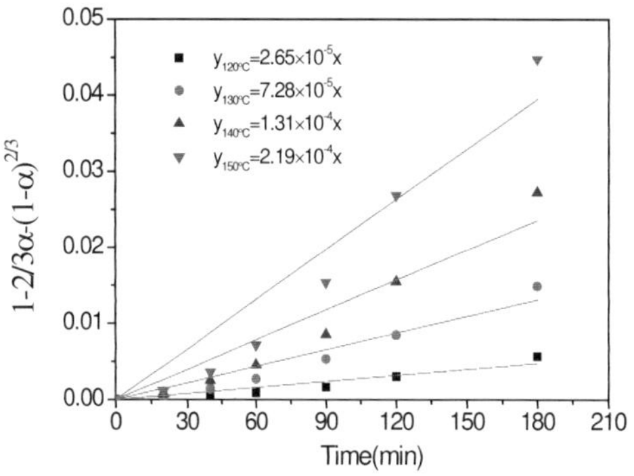

Figure.4. Relationship of temperature $1-2\alpha/3-(1-\alpha)^{2/3}$ with time

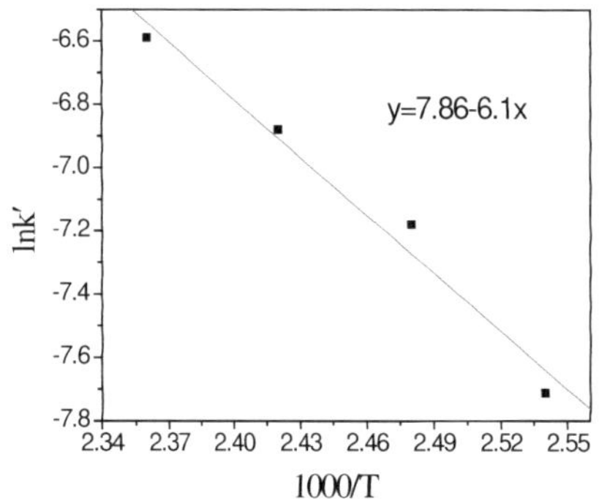

Figure.5. Arrhenius plot of chemical reaction control

Fig. 3 the straight slope of the logarithm of different temperature. The results are shown in Fig. 5. According to Arrhenius theorem [15]:

$$d\ln K'/dT = Ea/RT^2$$

The integration is as follows:

$\ln K' = -Ea/RT + C$

The apparent activation energy of chemical reaction control is 50.7kJ/mol according to Fig.6. According to the shrinking core model kinetic theory, when the apparent activation energy is 8~20 kJ/mol, the leaching process is internally diffusions control. When the apparent activation energy is over 40 kJ/mol, the leaching process is chemical reaction control. It can be estimated that the leaching process is chemical reaction control.

We could set up the kinetic equation which shows the temperature's effect on copper leaching according to the result in Fig. 5.

$$\ln K' = 7.86 - 6.1\times10^3 \times 1/T \quad (7)$$

3.3. The effect of particle size on copper leaching efficiency.

The sulfuric acid is 50 g/L, pressure 1.0 MPa, temperature 130 °C, stirring rate 600 rpm, a small amount of lignin is added, the effect of particle size on the copper leaching rate shown in fig. 6.

We work out the relationship between different $1-(1-\alpha)^{1/3}$ of particle size and time. The result is shown in Fig. 7.

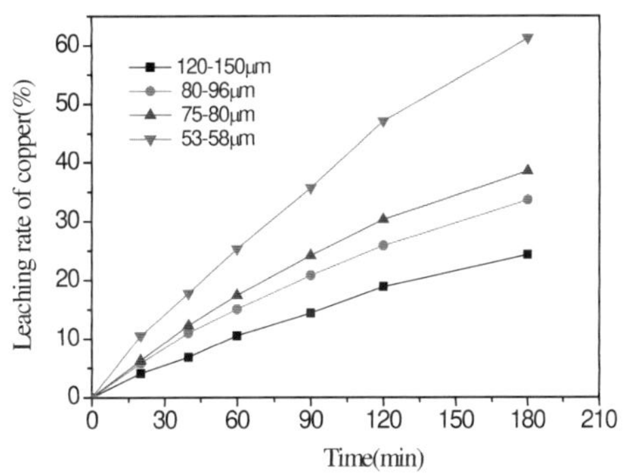

Figure. 6. Effect of mineral grain-size

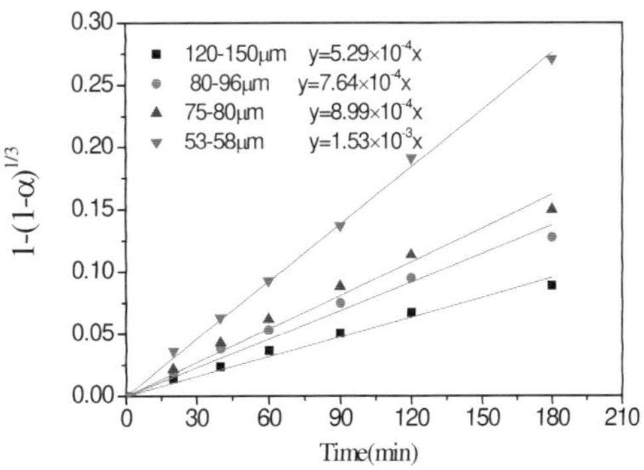

Figure. 7. Relationship between different $1-(1-\alpha)^{1/3}$ of particle size and time

According to Fig. 7, we could work out the relationship between ln K'and $\ln r_0$. The result is shown in Fig. 8.

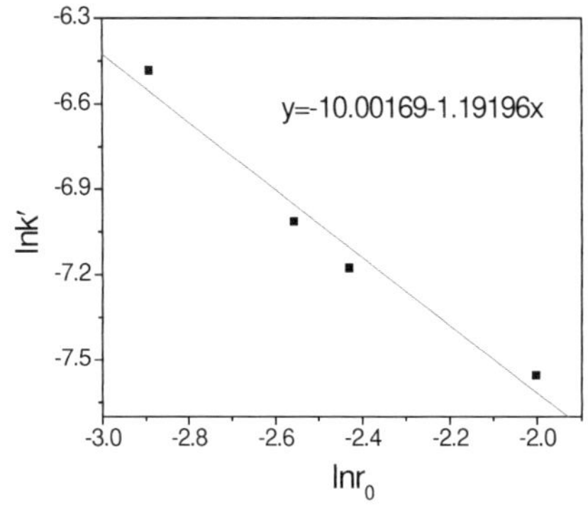

Figure. 8. The relationship ln K′with lnr$_0$

We could set up the kinetic equation which shows the particle size's effect on copper leaching according to the result in Fig. 8.

$$\ln K' = -10.00169 - 1.19196 \times \ln r_0 \tag{8}$$

3.4. The effect of pressure on copper leaching efficiency.

Particle size 80~96 μm, the sulfuric acid 50 g/L, temperature 130 °C, stirring rate 600 rpm, a small amount of lignin is added. Test the effect on copper leaching rate when the pressure is within the range of 0.8~1.4 Mpa.

We work out the relationship between different $1-(1-\alpha)^{1/3}$ of pressure and time. The result is shown in Fig. 10.

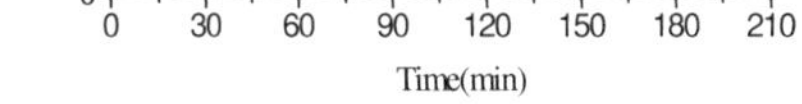

Figure. 9. Effect of pressure

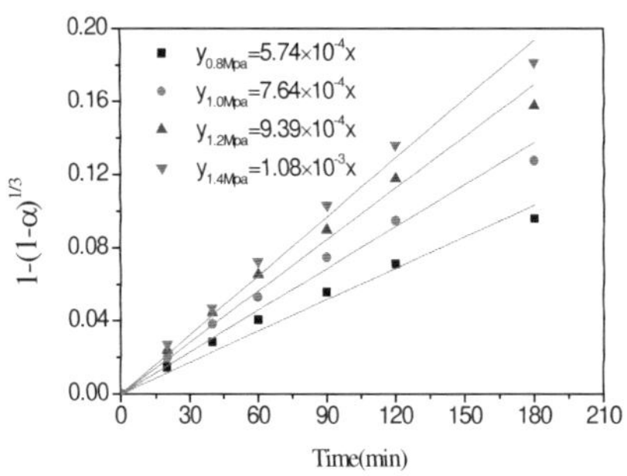

Figure. 10. Relationship of pressure $1-(1-\alpha)^{1/3}$ with time

According to Fig. 10, we could work out the relationship between ln K′and lnP. The result is shown in Fig. 11.

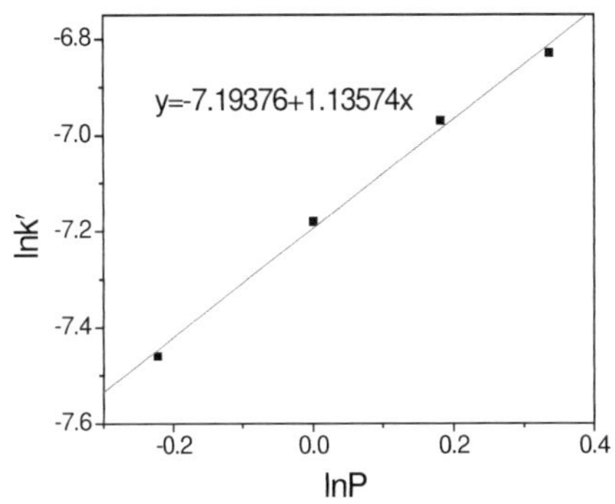

Figure. 11. Relationship ln K′with lnP

We could set up the kinetic equation which shows the pressure's effect on copper leaching according to the result in Fig. 11.

$$\ln K' = -7.19376 - 1.13574 \times \ln P \tag{9}$$

3.5. The effect of initial acidity on copper leaching efficiency.

Particle size 80~96 μm, pressure 1.0 MPa, temperature 130 °C, stirring rate 600 rpm, a small amount of lignin is added, the effect of initial acidity on copper leaching rate shown in fig.12.

We work out the relationship between different $1-(1-\alpha)^{1/3}$ of sulfuric acid and time. The result is shown in Fig. 13.

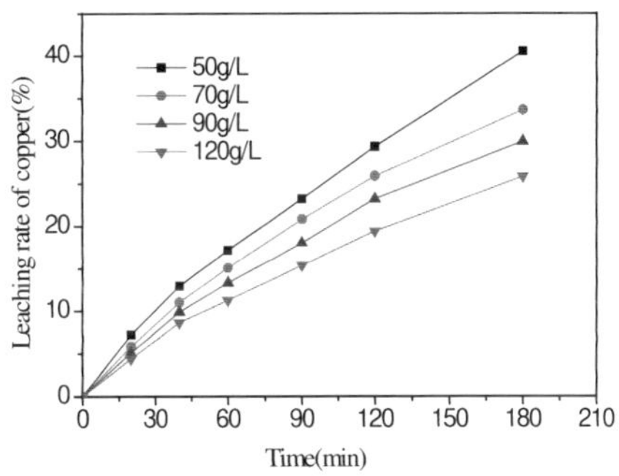

Figure. 12. Effect of sulfuric acid

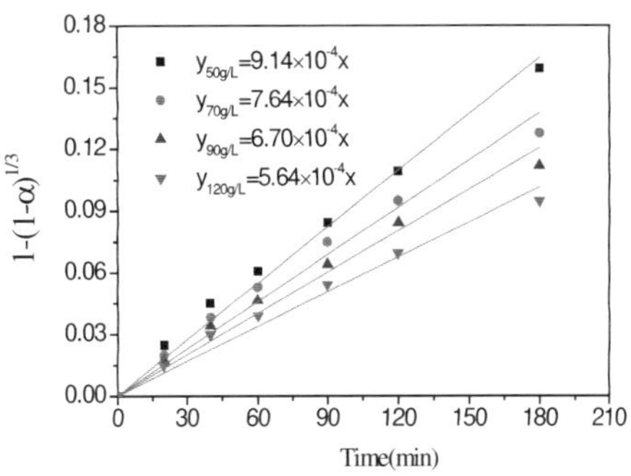

Figure. 13. Relationship between different $1-(1-\alpha)^{1/3}$ of sulfuric acid and time

According to Fig. 13, we could work out the relationship between ln K′acid and ln C. The result is shown in Fig. 14.

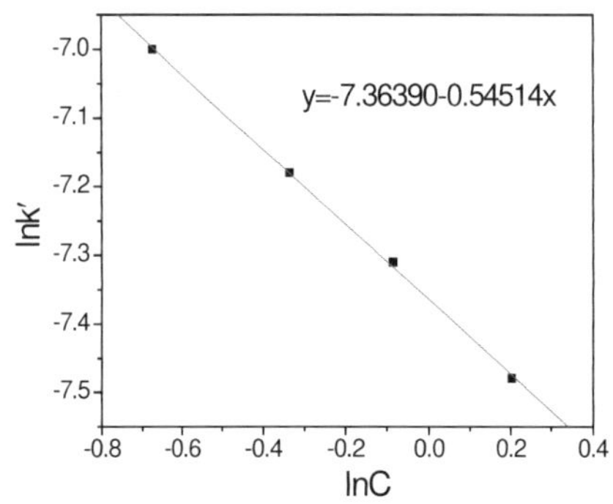

Figure. 14. Relationship ln K'with lnP

We could set up the kinetic equation which shows the sulfuric acid's effect on copper leaching according to the result in Fig. 14.

$$\ln K' = -7.36390 - 0.54514 \times \ln C \tag{10}$$

4. Establish Kinetic equation

The test above shows that the equations (7),(8),(9),(10) are all linear relationship. Multiple linear regression analysis is conducted on dependent variable ln k'and independent variable 1/T、 ln P、 ln C、 ln r_0 with SPSS software. The regression equation is obtained:

$$\ln K' = 3.624 - 5629.521\frac{1}{T} + 1.120\ln P - 0.542\ln C - 1.223\ln r_0$$

Therefore, we could work out the copper leaching kinetic equation in the process of pressure acid leaching of sulphide ore mixed with copper and zinc .

$$\alpha = 1 - \left[1 - t \times \exp\left(3.624 - 5629.521\frac{1}{T} + 1.120\ln P - 0.542\ln C - 1.223\ln r_0\right)\right]^3$$

5. Conclusions

(1) The test using 2 L autoclave to simulate practical production.When the stirring rate is over 600 rpm, it almost has no effect on copper leaching rate. Therefore under this condition, the external diffusions process is not the copper leaching reaction controlling steps.

(2) In the process of pressure acid leaching of sulphide ore mixed with copper and zinc, the apparent activation energy of copper leaching chemical reaction control is 50.7 kJ/mol. We could determine that the copper leaching reaction is shrinking core model controlled by chemical reaction.

(3) The influence of temperature, initial acidity, pressure, particle size on copper leaching. The kinetic equation of copper leaching in the process of pressure acid leaching of sulphide ore mixed with copper and zinc is listed as follows:

$$\alpha = 1 - \left[1 - t \times \exp\left(3.624 - 5629.521\frac{1}{T} + 1.120\ln P - 0.542\ln C - 1.223\ln r_0\right)\right]^3$$

(4) In fact, this kinetic equation of copper has deviation with practical production. This result occurred because the stirring rate is lower than 600 rpm and the initial acidity value change with time in practical production.

IOP Conf. Series: Materials Science and Engineering **303** (2017) 012022 doi:10.1088/1757-899X/303/1/012022

References

[1] Xie K Q, Yang X W, Shu Y Z and Shen Q F 2006 China Nonferrous Metallurgy vol 4 p 19
[2] F Habashi 2004 The Canadian Institute of Mining, Metallurgy and Petroleum p 3
[3] F Habashi 1997 Proc. Of the XX IMPC (Aachen) vol 1 p 129
[4] Mike Anthony and Doug Flett 2000 England Mineral Industrial Research Organization vol 2 p 55
[5] Misra M and Fuerstenau M. C 2005 Minerals Engineering vol 18 p 293
[6] Havlik T, Skrobian M, Balaz P and et al 1995 INT J M INER PROCESS vol 43 p 61
[7] Mahajan V, Misra M, Zhong K and et al 2007 Minerals Engineering vol 20 p 670
[8] Liu Z X, Yin Z L, Hu H P and et al 2012 J. Cent. South Univ vol 19 p 77
[9] Xing W G, Zhong Z Q, Mei G G 1990 Nonferrous Metals(Extractive Metallurgy) vol 5 p 27
[10] McDonald R G, Muir D M 2007 Hydrometallurgy vol 86 p 191
[11] Jin B J, Yang X W, Shen Q F 2009 Hydrometallurgy vol 99 p 119
[12] FAN X X, PENG J H, HUANG M Y and et al 2005 Precious Metals vol 26 p 15
[13] Arslan F, Bulut G, Kangal M O and et al 2004 Scandinavian Journal of Metallurgy vol 33 p 6
[14] Qiu T, Nie G, Wang I and et al 2007 Trans Nonferrous Met Soc China vol 17 p 418
[15] Hua Y X 2004 An Introduction to Kinetics of metallurgical processes (Bei Jing: Metallurgical Industry Press) p 188-195
[16] Mo D C 1987 Kinetics of metallurgical (Chang Sha: Central south university press) p 173

IOP Conf. Series: Materials Science and Engineering **303** (2017) 012023 doi:10.1088/1757-899X/303/1/012023

Ternary Blends of High Aluminate Cement, Fly ash and Blast-furnace slag for Sewerage Lining Mortar

L C Chao[1], **C P Kuo**[2,*]

[1,2] National Kaohsiung First University of Science and Technology, Kaohsiung 824, Taiwan

*u0315914@nkfust.edu.tw

Abstract. High aluminate cement (HAC), fly ash (FA) and blast-furnace slag (BFS) have been treated sustainable materials for the use of cement products for wastewater infrastructure due to their capabilities of corrosion resistance. The purpose of this study is to optimize a ternary blend of above mentioned materials for a special type of mortar for sewerage lining. By the using of Taguchi method, four control parameters including water/cementitious material ratio, mix water content, fly ash content and blast-furnace slag content were considered in nine trial mix designs in this study. By evaluating target properties including (1) maximization of compressive strength, (2) maximization of electricity resistance and (3) minimization of water absorption rate, the best possible levels for each control parameter were determined and the optimal mix proportions were verified. Through the implementation of the study, a practical and completed idea for designing corrosion resistive mortar comprising HAC, FA and BSF is provided.

1. Introduction

Infrastructure such as concrete sewer pipes, wastewater collection units and treatment plants are facing severe concrete cracks, fractures, deformations and collapse due to the sulfuric acid attack produced by sulfur oxidizing bacteria [1]. Therefore, it is critical to increase the durability of concrete which can reduce the biogenic acid attack [2] or to apply effective coating (lining) to concrete to protect the cement pipe surface from the acid attack [3]. With several distinct advantages over OPC concrete, HAC has been used in refractories, pipes, sewer constructions, industrial floors and dam spillways and more recently immobilization of metallic radioactive wastes and toxic metal solutions [4]. Research also indicated that HAC can be blended with different additions such as FA, BFS or natural pozzolans to obtain properties such as fast setting, rapid hardening, high early strength and shrinkage compensation [5]. This study, through the implementation of Taguchi's experiment design method, reports a study of blends that composing HAC, FA and BFS to develop the optimal mortar mix design parameters for sewerage lining in respect to the maximization of compressive strength, electricity resistance and the minimization of water absorption rate.

2. Experimental significance

Various researches have been done on improving performance of concrete used in sewerage. In which, using alumina cement, incorporating supplementary cementitious materials (SCMs), designing low water-to-cement (w/c) ratio concrete, applying concrete with low water absorption rates and increasing the concrete electrical resistivity are primary methods to increase concrete durability and resistance to against sulfate attack [4,5,6,7]. Concrete compressive strength is regarded as a valuable index in concrete structural design because it associates with most other properties and provides an overall quality parameter [8,9]. Electrical resistivity is an evaluation of the accessibility of aggressive agents before the beginning of the corrosive process [10], and is therefore used as a measurement of concrete durability [11]. Water absorption of concrete can be described as the ability to take in water or other aggressive agents by means of capillary action. The lower the water absorption in concrete, the finer the concrete is. It is treated an important factor for quantifying the concrete durability. [12,13]. Based on the research results aforementioned, criteria shown below were designed to aim and comply in this study:

- W/cm % for each mix was assigned to be less than 0.4;
- Compressive strength tests performed according to ASTM C109/C109M [14], must be greater than 210 kgf/cm^2;
- Electricity resistance tests performed according to AASHTO T277 [15], must be greater than 20 kΩ-cm;
- Water absorption tests performed according to the ASTM C 642 [16], must be less than 8.0 %.

3. Experimental details

3.1. Materials

HAC used met the requirements of GB201-2000 [17] with a specific gravity of 3.30 and a blaine fineness of 3700 cm^2/g. FA and BFS used had properties that met the requirements specified in ASTM C618 [18]. The FA had a specific gravity of 2.69 and blaine fineness of 4080 cm^2/g and the BFS had a specific gravity of 2.90 and blaine fineness of 4300 cm^2/g. Standard sand with properties complied with ATM C778 [19] was used as fine aggregate. Tap water was used as mix water for each mix.

3.2. Taguchi method

Taguchi method was developed by Genichi Taguchi during the 1950s to be used to determine optimal parameters for performance and cost [20]. The method recommends the use of "loss function" to measure the deviation of obtained properties from the target properties. The value of loss function is further transformed into signal-to-noise (S/N) ratio. There are three performance characteristics to analyze the S/N:

$$S/N = -10 * \log_{10}\left(\frac{1}{n}\sum_{i=1}^{n}(Yi - Y_0)^2\right) \quad (1)$$

- Larger is better: The S/N can be calculated using Eq. (2).

$$S/N = -10 * \log_{10}\left(\frac{1}{n}\sum_{i=1}^{n}\frac{1}{Y_i^2}\right) \quad (2)$$

- Smaller is better: The S/N can be calculated using Eq. (3).

$$S/N = -10 * \log_{10}\left(\frac{1}{n}\sum_{i=1}^{n}Y_i^2\right) \quad (3)$$

where n is the number of experiments and Yi is the observed results for test.

In this study, "the larger the better" situation was applied for maximization of compressive strength and electricity resistance and "the smaller the better" situation was applied for minimization of water absorption rate.

3.3. Mix design using Taguchi method

Four control parameters including water/cementitious material ratio (W/cm), mix water content (W), fly ash content (FA/cm) and blast-furnace slag content (BFS/cm) at 3 levels were considered (See

Table 1) in this study. According to the Taguchi method (standard L_9 orthogonal array for four control parameters at 3 levels), the suggested experiments and the detailed mix proportion of the nine experiments were designed (see Tables 2).

Table 1. Control parameters and their levels.

parameter 1 W/cm (%)	parameter 2 W (kg/m3)	parameter 3 FA/cm (%)	parameter 4 BFS/cm (%)
35	140	7.5	7.5
37	160	10	10
39	180	12.5	12.5

Table 2. Taguchi Standard L9 orthogonal array and detailed mix proportion.

Trial No.	Performance parameter value							
	parameter 1	parameter 2	parameter 3	parameter 4	W/cm (%)	W (kg/m^3)	FA/cm (%)	BFS/cm (%)
M1	1	1	1	1	35	140	7.5	7.5
M2	1	2	2	2	35	160	10	10
M3	1	3	3	3	35	180	12.5	12.5
M4	2	1	2	3	37	140	10	12.5
M5	2	2	3	1	37	160	12.5	7.5
M6	2	3	1	2	37	180	7.5	10
M7	3	1	3	2	39	140	12.5	10
M8	3	2	1	3	39	160	7.5	12.5
M9	3	3	2	1	39	180	10	7.5

3.4. Test procedures and results

A 0.1 m^3 laboratory twin-shaft mixer manual electric mixer was used for mortar mixes. The total mix time was four minutes for each mix. Three 100mm x 100 mm cubic specimens for compressive strength, three 100mm x 100 mm cubic specimens for water absorption and for electricity resistance test were collected for each trial (electricity resistance tests were conducted on the same specimens prepared for water absorption rate tests). Specimens were kept in molds in ambient condition for 24 hours after their castings and were then demolded and stored in ambient condition until testing. Tests were performed on specimens at age of 28 days. Results of tests of each mix were reported as averages of three specimens (See Table 3). Compressive strengths were in the range of 204-326 kgf/cm^2 and the highest value was measured at M7. Trial No. 9 failed in the compressive strength test since the result was not complied with the criteria. The electricity resistance test results were all above the criteria and were in the range of 41-224 kΩ-cm. The highest electricity resistance was measured at M7. The results of water absorption rate test were in the range of 4.8-6.8 % and were all complied with criteria. The lowest water absorption rate was measured at M7, which was the best.

Table 3. Test results of trials.

Trial No.	Compressive strength, (kgf/cm^2)	Electricity resistance, (kΩ-cm)	Water absorption rate, (%)
M1	266	103	5.0
M2	289	138	6.6
M3	275	56	5.7
M4	230	41	6.8
M5	234	52	6.2
M6	269	84	6.4
M7	326	224	4.8
M8	272	159	6.8
M9	204	96	7.1

4. Optimizing mix design proportions

The situations of "the larger the better" were used for maximization of compressive strength and electricity resistance and situations of "the smaller the better" were used for minimization of water absorption rate in this study. The S/N ratios of compressive strength and electricity resistance obtained by using Eq. (2) and water absorption rate obtained by using Eq. (3) (See Table 4). Based on the Taguchi method, the response S/N for each level of a parameter was obtained by averaging the S/Ns of that respective level (See Table 5). A higher response S/N value indicates a greater effect of the particular control parameter at that level. The proposed optimal mix proportions should be determined (See Table 6).

Table 4. S/N of results of trials.

Trial No.	Parameter levels				S/N		
	W/cm (%)	W (kg/m^3)	FA/cm (%)	BFS/cm (%)	Compressive strength	Electricity resistance	Water Absorption rate
M1	35	140	7.5	7.5	47.6	40.3	-14
M2	35	160	10	10	48.7	42.8	-16.4
M3	35	180	12.5	12.5	48.5	35	-15
M4	37	140	10	12.5	47	32.3	-16.6
M5	37	160	12.5	7.5	16.3	34.3	-15.8
M6	37	180	7.5	10	47.6	38.5	-16.2
M7	39	140	12.5	10	49.4	47	-13.6
M8	39	160	7.5	12.5	46.7	44	-16.6
M9	39	180	10	7.5	46.6	39.6	-17

Table 5. Response S/N for each level of a parameter on properties.

Properties	W/cm (%)			W (%)			FA/cm (%)			BFS/cm (%)		
	35	37	39	140	160	180	7.5	10	12.5	7.5	10	12.5
Compressive strength	48.8	47.7	48.4	47.7	48	48.8	48.6	47.5	48.8	47.4	49.4	48.2
Electricity resistance	39.3	35	43.6	39.5	40.3	38.1	40.9	38.2	38.8	38.1	42.8	37.1
Water absorption rate	-15.2	-16.2	-15.8	-15.5	-16.5	-15.1	-15.6	-16.7	-14.8	-15.6	-15.4	-16.1

Table 6. Proposed optimal mix proportions.

Optimal trial No.	Optimal mix proportions	W/cm (%)	W (kg/m3)	FA/cm (%)	BFS/cm (%)
1	Compressive strength (kg/cm^2) / Water absorption rate (%)	35	180	12.5	10
2	Electricity resistance (kΩ-cm)	39	160	7.5	10

5. Verifying the optimal mix proportions

Since the proposed optimal mix proportions (see Table 6) were not included in the mix proportions (see Table 2), verification trials were performed. The tests were performed on specimens at age of 28 days, experiment results showed that the proposed optimal mix proportions satisfied the expectancies (see Table 7).

Table 7. Optimal mix design verification test results (average of three specimens).

Target properties	Optimal trial No.	
	1	2
Compressive strength (kg/cm2)	293	286
Electricity resistance (kΩ-cm)	59	162
Water absorption rate (%)	4.3	5.4

IOP Conf. Series: Materials Science and Engineering **303** (2017) 012023 doi:10.1088/1757-899X/303/1/012023

6. Conclusions

As an effective and suitable experiment method for optimizing the proportion parameters of ternary blends of HAC, FA and BFS, Taguchi method was used to optimize the properties of sewerage lining mortar, following conclusions can be drawn from the results: Four control parameters including water/cementitious material ratio (W/cm), mix water content (W), fly ash content (FA/cm) and blast-furnace slag content (BFS/cm) were used in nine trial mix designs to maximize the compressive strength and electricity resistance and to minimize of water absorption. Two set of mix designs were verified optimal. Results of target properties of each optimal mix designs complied with the criteria.

7. References

[1] Aderhold J, Davydov V Yu, Fedler F, Klausing H, Mistele D, Rotter T, Semchinova O, Stemmer J and Graul J 2001 *J. Cryst. Growth* **222** 701
[2] [1] Pacheco-Torgal F, Abdollahnejad Z, Miraldo S, Baklouti S and Ding Y 2012 *Constr. Build.Mater.* **36** 1053-1058
[3] Yuan H, Dangla P, and Chatellier P 2015 *Cem. Concr. Res.* **70** 29-38
[4] Chindaprasirt P and Rattanasak U 2016 *Constr. Build.Mater.* **112** 956-961
[5] Vafaei M and Allahverdi A 2016 *Constr. Build.Mater.* **114** 290-296
[6] Kinoshita H, Swift P, Utton C, Carro-Mateo B, Marchand G, Collier N and Milestone N 2013 *Cem. Concr. Res.* **50** 11-18
[7] Araghi H, Nikbin I, Reskati S, Rahmani E and Allahyari H 2015 *Constr. Build.Mater.***77** 461-471
[8] Rostami V, Shao Y and Boyd A 2011 *Constr. Build.Mater.* **25** 3345-3355
[9] Ishida T, Luan Y, Sagawa T and Nawa T 2011 *Cem. Concr. Res.* **41** 1357-1367
[10] Behnood A, Verian K and Gharehveran M 2015 *Constr. Build.Mater.* **98** 519-529
[11] Hunkeler F1996 *Constr. Build.Mater.* **10** (5) 381-389
[12] Lubeck A, Gastaldini A, Barin D and Siqueira J 2012 *Cem. & Concr. Compos.* **34** 392-399
[13] Kirca O, Yaman I and Tokyay M 2013 *Cem. & Concr. Compos.* **35** 163-170
[14] Hornbostel K, Larsen C, Geiker M 2013 *Cem. & Concr. Compos.* **39** 60-72
[15] ASTM 2017 *Standard Test Method for Compressive Strength of Hydraulic Cement Mortars (Using 2-in. or [50-mm] Cube Specimens) ASTM C109 / C109M - 16a* West Conshohocken PA
[16] AASHTO 2015 *Standard test method for electrical indication of concrete's ability to resist chloride ion penetration AASHTO T277-15* Washington D.C.
[17] ASTM 2013 *Standard Test Method for Density, Absorption, and Voids in Hardened Concrete ASTM C642-13* West Conshohocken PA
[18] CSBTS 2000 *Aluminate cements GB201-2000* Beijing China
[19] ASTM 2015 *Standard Specification for Coal Fly Ash and Raw or Calcined Natural Pozzolan for Use in Concrete ASTM C-618-15* West Conshohocken PA
[20] ASTM 2013 *Standard Specification for Standard Sand ASTM-C778/13* West Conshohocken PA
[21] Ranjit K1990 *A primer on the Taguchi method* Van Nostrand Reinhold New York